# Tourism in Egypt
# Through the Ages

# Tourism in Egypt Through the Ages

## A Historical Guide

Charlotte Booth

First published in Great Britain in 2024 by
Pen & Sword History
An imprint of Pen & Sword Books Limited
Yorkshire – Philadelphia

Copyright © Charlotte Booth 2024

ISBN 978 1 39904 356 4

The right of Charlotte Booth to be identified as
Author of this Work has been asserted by her in accordance
with the Copyright, Designs and Patents Act 1988.

A CIP catalogue record for this book is
available from the British Library

All rights reserved. No part of this book may be reproduced or
transmitted in any form or by any means, electronic or mechanical
including photocopying, recording or by any information storage and
retrieval system, without permission from the Publisher in writing.

Typeset by Mac Style
Printed in the UK by CPI Group (UK) Ltd, Croydon, CR0 4YY.

Pen & Sword Books Limited incorporates the imprints of After
the Battle, Atlas, Archaeology, Aviation, Discovery, Family History,
Fiction, History, Maritime, Military, Military Classics, Politics,
Select, Transport, True Crime, Air World, Frontline Publishing, Leo
Cooper, Remember When, Seaforth Publishing, The Praetorian Press,
Wharncliffe Local History, Wharncliffe Transport, Wharncliffe True
Crime and White Owl.

For a complete list of Pen & Sword titles please contact

**PEN & SWORD BOOKS LIMITED**
47 Church Street, Barnsley, South Yorkshire, S70 2AS, England
E-mail: enquiries@pen-and-sword.co.uk
Website: www.pen-and-sword.co.uk
or
**PEN AND SWORD BOOKS**
1950 Lawrence Rd, Havertown, PA 19083, USA
E-mail: uspen-and-sword@casematepublishers.com
Website: www.penandswordbooks.com

# Contents

*Acknowledgments*     vi
*Introduction*     vii

**Chapter 1**     The Importance of Tourism in the Twentieth Century     1

**Chapter 2**     The Beginnings of the Modern Tourism Industry     26

**Chapter 3**     Travel in the Seventeenth and Eighteenth Centuries     55

**Chapter 4**     Christian Pilgrimages to Egypt     71

**Chapter 5**     Medieval Tourism     89

**Chapter 6**     Graeco-Roman Tourism     107

**Chapter 7**     Visiting the Oracle     134

**Chapter 8**     Pharaonic Travel     158

*Epilogue: The Future of Tourism in Egypt*     179
*Notes*     185
*References*     198
*Index*     203

# Acknowledgments

Whilst writing a book is primarily a solitary activity, there are a team of people who help to make it possible.

I would like to thank the Society of Authors, who through their Authors' Foundation Grant helped to finance some of the research trips needed. Thanks also go to the editors at Pen and Sword who were patient when during lockdown this book had to go on hold as all the libraries were closed to researchers.

Of course special thanks go to Billy, who ensured I shut myself away to write, bringing me tea and not allowing me to get distracted.

# Introduction

In March 2020, all international borders were closed, and travel became another thing that we were unable to do due to the global pandemic. Of all the restrictions this was the one I personally struggled with the most. Not being able to travel to new and interesting places, learn about new cultures and walk in the footsteps of history was difficult. I know I was not the only one feeling this loss, as travel is an important part of modern Western culture. When I started considering my reasons for travelling, and what I enjoy about it, I wondered whether the ancient Egyptians also loved to travel around their own country, were interested in the monuments of their ancestors or enjoyed seeing new and interesting things. Through my research over the years there have been many clues that the ancients did stray from their villages, travelled on business and visited family, but it got me wondering whether they had 'tourism' as we understand it? It was these ponderings which led to this book.

In order to answer the question of whether the ancient Egyptians had 'tourism', we need to define the term 'tourism' before we can hope to apply it to those living in the Nile Valley 5,000 years ago.

According to the *Cambridge Dictionary* a tourist is:

> someone who visits a place for pleasure and interest, usually while on holiday.[1]

And tourism is:

> the business of providing services such as transport, places to stay, or entertainment for people who are on holiday.

It is important to define the word 'holiday' as well:

> a time when someone does not go to work or school but is free to do what they want, such as travel or relax.

Could this mean that a weekend is essentially a holiday, and if you go 'sightseeing' at this time, does this make you a 'tourist'? According to these definitions I would say yes. Through the pandemic there was a greater importance attached to being a tourist in your hometown in the absence of foreign travel. Holidays are therefore, by default, often connected with the concepts of tourists and tourism. However, there is a group who define themselves as 'travellers' rather than 'tourists', even though they are not working in the country they are staying in and are essentially on an extended 'holiday'. So, what is the difference? In her PhD thesis, Victoria Foertmeyer, suggests very little:

> travellers are defined as tourists if they have travelled to a site for the purpose of sightseeing, or if they are making a pilgrimage or pursuing some other quests, even in the course of employment or business.[2]

She goes further by defining anyone as a tourist who describes how they feel about a site and the impressions they have of it – whether this be in a travel guide, a travel journal or a blog. In this book, the term 'tourist' therefore describes people travelling on pilgrimage, royal visits to new temples and towns, people travelling on business and those who are travelling purely for recreation.

In the grand scheme of things such semantics are not important, but in recent decades the term 'tourist' is viewed almost as derogatory, with many preferring to be described as a 'traveller', even though essentially they are the same thing. J.L. Beness and T. Hillard, for example, writing about earliest Latin graffiti in Egypt ask, 'Were these merely the graffiti of tourists?',[3] as if this would diminish the importance and significance of it as a historical document. Although considered important today, such differences in semantics, simply did not exist in ancient Egypt. In fact, they did not have a word for traveller or tourist at all. But that does not mean that people did not travel.

Tourism and the tourism industry as we understand it in the modern world, as a multi-million-pound industry, did not come into existence until the nineteenth century. It could be argued, however, that in the Roman period and earlier, entrepreneurial spirits were making money out of tourists, seeing the start of a rudimentary tourist industry (chapter 6). This goes to show that even without a fully-fledged industry in place people still travelled, and others made money out of facilitating the journey. The task for this book is to trace this trajectory back beyond the nineteenth century, and beyond the Roman era to the pharaonic era.

## Hierarchy of Travel

Research has shown that throughout the history of travel in Egypt, before and after the introduction of a tourism industry, travellers were characterised depending on their purpose for travelling. From the eighteenth century onwards for example, there were two groups of people who travelled to Egypt: archaeologists and tourists. The latter group included leisure travellers, artists, businessmen, health tourists and pilgrims, and all came from the same upper-middle or upper classes of Western society.[4] Travelling at this time was expensive and was viewed as a means of improving status in society through increasing knowledge and broadening horizons.[5] It was considered by many to be a 'populist form of respectability'.[6] Travelling was a means of improving social standing.

Despite all coming from the same social class there was a hierarchy of travellers with archaeologists feeling they were superior to general tourists. However, many archaeologists started out as tourists, as a large proportion of visitors went to Egypt in order to purchase or excavate monuments and antiquities.[7] Many eighteenth- and early nineteenth- century travellers were involved to some extent in 'exploration, anthropology, survey, collection, excavation, plunder, theorizing and prophecy'.[8] However, once an archaeologist received an official concession to excavate, they considered 'tourists' to be as annoying as flies and resented their presence.

Archaeologists of the nineteenth century further classified 'bad travellers' as those who were unscientific and lacked certain knowledge; essentially tourists, whereas 'good travellers' were scientific, systematic record keepers (i.e. archaeologists and Egyptologists).[9] This in itself is problematic as not all archaeologists were trained or scientific in their approach to the antiquities of Egypt.

For example, Montagu Ballard (1850–1936), an excavator at Giza in 1901–2 was considered 'bad' as he 'left a horrific path of destruction as he ripped his way through the large cemetery, revealing its previously unsuspected riches to tomb robbers'.[10] By modern archaeological standards many of the archaeologists and 'good travellers' of the nineteenth and early twentieth centuries were 'bad' as they caused a lot of damage to the monuments. Amelia Edwards (1831–1892) for example carved her name onto the walls of Abu Simbel and James Henry Breasted (1865–1935) was sacked from his first Nubian excavation for carving his name onto the temple wall.[11] Howard Carter, who found the tomb of Tutankhamun, decapitated the boy king in order to remove the golden death mask, as well as leaving the mummy out in 50-degree desert heat to try to melt some of the resin used in the mummification process to make removal of the jewellery easier.

Despite these (and countless similar failings) archaeologists' training (no matter how rudimentary) and status rendered their actions acceptable and almost beyond reproach, whereas tourists doing the same were condemned by the professional archaeologists.

Others in the eighteenth and early nineteenth centuries believed pilgrims to be the superior visitors in the hierarchy as they were travelling for spiritual reasons rather than frivolous leisure. The diary of one pilgrim, Father Dolák (1870) was heavily criticised as he appeared to be enjoying himself too much for a pilgrim, and behaved like an amused tourist, 'with all the burden this term carries'.[12]

This theme had been debated for more than a century with Dean A. Walker in 1891 stating:

> the ordinary tourist is to be pitied. He sees the principal places when tired from a hard day's ride. He has not the language, and for information must depend on the '*Baedeker*' which is now on many points out of date, and on his dragoman, who thinks he is not earning his pound a day if his stories fall below the maximum size.[13]

J. Buzard adds that being a tourist in Egypt was predictable and repetitive, whereas 'travellers' wanted to discover the 'real' Egypt, acting spontaneously and travelling off the beaten track in an exotic and unpredictable way.[14] This desire for a 'real' experience is what apparently separates travellers from tourists even in the modern world. However, can you ever discover the 'real' Egypt as an outsider? If you do not live there permanently? If you do not speak the language fluently, and have not the same cultural beliefs? Can you experience the life of the Egyptians when from a middle/upper class Western background?

These questions were just as valid two centuries ago as they are today, and many believed to be a good traveller it was necessary to abandon Western luxuries and live more frugally. Giovanni Belzoni said of Henry William Beechey (c. 1789–1862) that:

> after having weaned himself from those indulgences to which he was accustomed ... would make a good traveller.[15]

It seems quite a strange concept that since the nineteenth century visitors to Egypt want to differentiate themselves from 'tourists' by claiming they were 'travelling' or 'here on business', staying in apartments or more recently that they are 'digital nomads'. However, it all amounts to the same thing. If you are visiting a place for the first time and enjoying cultural experiences or visiting a monument, you are a tourist, even if you are in the office during the day or staying in a hostel for three months. Additionally, following V.A. Foertmeyer's definition that if you write about your experiences this indicates you are a tourist, creates a catch-all for the travel blogs, where 'backpackers' and 'travellers' write about their experiences ... as tourists.

It is in this vein that I am using the words 'tourism' and 'tourist' throughout this book – regardless of how individuals wished to be disassociated with the term. The term tourism itself is such a new concept, I am using it to mean 'to travel some distance from home to visit new places' regardless of the reason, whether business, pilgrimage, or pleasure.

## Practicality of Travel

Without getting caught up too much in the snobbery of semantics, it is important to consider the difficulty in travelling what we would now consider short distances throughout most of the historical period. Even travelling to another town was considered an ordeal – which goes some way to explaining why many people recorded (and still record) their trip in the form of diaries, graffiti and letters. It is these 5,000 years of journeys that I hope to record in this book to show travel may have been as important to the ancient Egyptians as it is to us today. Their journeys may not have been as frequent or as easy, and were often associated with work or religion rather than pleasure, but they were important.

Part of the increase in travel in the later nineteenth and early twentieth centuries was due to travel becoming easier, quicker and more accessible. This, however, would have been very different for the ancient Egyptians. Over a period of 5,000 years, the modes of transport evolved and developed alongside technological advances and imports from other countries. This meant that as time progressed transport became more convenient and reliable and, as a consequence, travel became more widespread.

There were two main forms of transport in Egypt throughout the majority of its history which comprised overland and river travel. A third was then added in the second half of the twentieth century when air travel became a reliable and accessible option. It was a game changer for the travel industry.

## Overland Travel

In any age the key to successful overland travel is a network of serviceable roads and in pharaonic Egypt this network was extensive but, to the larger extent, organic. For example, when an irrigation canal was built the displaced sand and earth formed embankments alongside the canal, which were then flattened by the passage of people and animals creating roads. The same thing happened with paths leading between villages and across deserts, where the path of least resistance was created and then repeatedly used. We can see the same form of organic path creation in the modern world where short cuts through a fence or down an embankment end up looking like a path, and therefore guide future users. Such paths and roads in ancient times were not paved and simply became impacted by repeated use. Some of these paths are still visible and still used.

It was only ceremonial pathways like the Avenue of the Sphinxes between Luxor and Karnak temple which were intentionally created and then paved in stone. However, not all paved roads were ceremonial, and a paved road has been discovered running from the basalt quarries at Gebel Qatrani in the Fayoum to Cairo. It was 11.5 kilometres long, and at the widest part it was 2.1 metres wide, and constructed with flagstones and petrified wood. It appears to have been built in the fifth and sixth dynasties (2498–2184 BCE),[16] and facilitated the transportation of basalt from the quarries to Lake Moeris where the journey was continued by water.

Desert paths were also created for recreational purposes as demonstrated by the road near Amenhotep III's (1386–1349 BCE) palace of Malkata on the west bank at Luxor. This seems to have been a racetrack for chariots or horses, which was maintained by clearing the rocks and flattening uneven surfaces. There was also a viewing platform which enabled onlookers to watch the race from an elevated position. It was important that roads for chariots were smooth, as even the sturdiest of vehicles could easily be damaged if riding on uneven ground. Whether they also had a road maintenance team who filled in potholes and generally kept general well-travelled roads passable for wheeled vehicles is unknown, but in theory it would not have been a particularly skilled job. Travellers themselves could simply pack any holes with desert material to allow safe passage and over time it would be compacted by use.

*Walking*
With rudimentary paths, or more maintained roads, travelling by foot was the only solution for many members of pharaonic society and may have limited the distance they could realistically travel; perhaps up to 16 kilometres a day. Most people in ancient Egypt were barefoot as sandals made of reeds and papyrus were expensive and, ironically, reserved for people who perhaps did not need to walk as they travelled by litter or donkey. Sandals cost between one and three deben of copper (91g), the same price as a goat.

Images of Egyptian soldiers show they marched barefoot for long distances across desert and rocky terrain, keeping in time and marching in step to the beat of a drum or a chant. In Middle Kingdom (2040–1782 BCE) models of army corps, such as those found in the tomb of Mesehti in Asyut, are all shown marching on the same leg.

At the other end of the scale, most people walked the short distance to work and within their local village or town but were unlikely to walk any great distance unless it was unavoidable.

*Donkey*

The donkey has been a staple of travel in Egypt since they were domesticated in the late pre-dynastic period (c. 2686 BCE). Although domesticated to be used as working animals, they were also used for long-distance overland travel. Essentially wherever a donkey could travel, so could a human, meaning even narrow, steep paths and landscapes were passable. There are very few images of people riding donkeys in the pharaonic tomb scenes, and it has been suggested that this could be due to the difficulties of representing a person astride a donkey.[17] However, there are scenes showing people riding donkeys whilst sat in a litter (a little wooden chair) on the back of the animals, but most images of donkeys show the drivers walking and the donkeys carrying goods. Donkeys can easily walk up to 25 kilometres a day whilst carrying heavy loads.

*Horses*

The horse was introduced to ancient Egypt at the beginning of the New Kingdom (1570 BCE). It is often stated they were introduced by the Hyksos kings, but the evidence seems to show that all the Near East saw the introduction at more or less the same time.

Ancient horses were smaller than modern horses and surviving chariot harnesses suggest they were about 13.5 hands (1.35 metres). The horse was introduced with the chariot and most representations are of horses pulling chariots rather than being ridden. From the representations that do exist, such as a New Kingdom model at the Metropolitan Museum of Art, New York (Acc 15.2.3), the rider is bareback without a saddle and just a simple bridle. Another image from the tomb of Horemheb at Saqqara shows a rider sitting on the horse at the back in the same way one would ride a donkey. The temple of Ramses III (1182–1151 BCE) at Medinet Habu also shows what appear to be soldiers in battle on horseback.

As for whether horses were used for travel, this is difficult to prove. It is possible, but they were an expensive commodity and only the very rich had access to them. However, by the Graeco-Roman period horseback riding was more common, but, again, the evidence does not survive to prove they were used for daily travel.[18]

*Chariots*

As mentioned, the chariot was introduced to Egypt at the same time as the horse and was used primarily by the wealthy for hunting or for the military. Two types of chariots were in use at the same time, the lighter Canaanite design, and the heavier Egyptian model.

Whilst the lighter one may have been practical as a form of general transportation, there is no surviving evidence demonstrating they were used as a means of travelling from A to B. The majority of temple and tomb scenes show the chariot in battle, being used for royal processions or in the desert for hunting; all of which are out of the scope of this book.

*Camels*

Camels are almost synonymous with Egypt and any modern tourist worth their salt has a picture of themselves on a camel in front of the pyramids at Giza.

However, camels were not introduced into Egypt until the Saite period (664 BCE) and by the Ptolemaic period (305–30 BCE) camels were being used as transport for long desert journeys. Saying that, however, evidence for the one-humped camel (dromedary) is known from the early dynastic period (3150–2686 BCE), but this was rare and they were not generally used for transportation.

## River Travel

As Herodotus famously said, Egypt was the 'land won by the Egyptians and given them by the Nile', and river travel was the number one means of travelling long and short distances from the pharaonic period until the introduction of the car and the train in the common era. Even in modern times it is quicker to go from the east to west bank at Luxor by boat than it is to go by road.

*Papyrus Skiffs*

Papyrus reed boats are the oldest to have sailed on the river Nile. Sadly none have survived due to the organic nature of papyrus but some of the earliest models of what are thought to represent papyrus skiffs date to the Badarian period (5,000 BCE).[19]

Prior to the New Kingdom (1570–1070 BCE) papyrus grew all along the Nile and was a useful material for various purposes, but after the New Kingdom it only grew in the Delta. Today it does not grow in the wild anywhere in Egypt and is cultivated in small amounts for the tourist market.

Whilst it is often assumed that papyrus skiffs were only used for short journeys, experiments in 1969 and 1970 show that they could be sailed across the sea and one of the reconstructed boats managed fifty-seven days across the Atlantic,[20] although this is unlikely to replicate the general usage of the ancient Egyptians.

These skiffs were often small and flat bottomed – essentially rafts – and were made by tying together bundles of papyrus reeds until the required tapering bow shape was achieved. Some may have had a small cabin on the deck to shelter from the sun or to store things needed for the trip. The deck was reinforced with wood to make a more stable platform to stand and sit on. Larger skiffs may have had masts, but smaller ones were probably steered with a long oar a little like a punt or a gondola.

Throughout the pharaonic period these skiffs were used for pleasure trips into the marshes in order to hunt fowl and to fish and this is depicted in numerous Middle and New Kingdom tombs, but they are also depicted being used by fishermen. They were able to support the weight of one or two people so would not have been suitable to transport heavy cargo.

Because the material grew freely along the Nile this type of vessel was available to anyone able to make one and with a lifespan of fifty-seven days they were a viable means of travelling the Nile on short trips or even the full length from Aswan to Cairo (880 kilometres).

Plutarch (first century CE) commented that the Egyptians believed that papyrus skiffs would not be attacked by crocodiles, which were a major threat in the Nile until the first Aswan Dam was built in 1899. There is possibly some truth to this as the skiff going through the water does not cause many ripples which are known to attract crocodiles.[21]

*Wooden boats*

For long journeys, royalty and the military, larger wooden boats were used instead of the lighter papyrus skiffs. A number of these boats have survived in the archaeological record, such as the funerary boat of Khufu (FIG 1) which was buried alongside the Great Pyramid in Giza. The earliest boat burial (albeit without the boat) was from the reign of Hor Aha (3050 BCE). In fact many of the boat pits discovered alongside tombs are empty as wood was an expensive commodity in a desert environment, and it was often plundered shortly after the burial. Many of the boats which were however discovered intact, such as the five boat burials of Senuseret III (1971–1926 BCE) at Dahshur are thought to be ceremonial as they do not appear to be watertight.

Many of the Old Kingdom images of wooden boats show they were similar in design to the royal boat of Khufu with a highly decorated cabin of some sort on deck with a mast which could be raised as needed. There were also a number of oars meaning it could be rowed. The sails were made of linen and were wider at the top than the bottom.

Larger vessels which were used for international travel and trade are depicted in various tombs and temples including the temple of Hatshepsut (1498–1483 BCE) at Deir el Bahri. Such vessels are really outside the scope of this book, but it should be known that they were not used for tourism within Egypt but rather international travel as part of military or trade routes.

However, it needs to be made clear that although we have hundreds of records of people travelling throughout Egypt, some mention they travelled by river but they rarely, if ever tell us the exact means of transport, and how this transportation was organised. We can assume that perhaps there were such things as 'water taxis' where boatmen charged a fee to transport people from east to west but also from north to south, but as these people were likely illiterate such records simply do not exist. We are, however, fortunate to have a letter from the third century CE which outlines some of the difficulties in transportation. Eutychis, in Antinoopolis wrote a letter to her mother Ametrion in Oxyrhynchus. She was trying to reach her mother but:

> I could not find any way to come to you, because the camel drivers refused to go to the Oxyrhynchite. Not only that, but I went up to Antinou(polis) for a boat and did not find any. So now I considered bringing my loads to Antinou and staying there until I find a boat to sail down.[22]

We will never know if Eutychis made the 140-kilometre journey, and if she did how she managed it, but the letter definitely arrived. We could wonder why she could not follow the same path as the messenger.

These modes of transport, as discussed, did not change much throughout the five millennia prior to the nineteenth century and the introduction of steamers, dahebeyas (traditional houseboats), cruise liners (see chapter 2) and of course cars, aeroplanes and coaches in the twentieth century. One thing that is increasingly frustrating is that the people who travelled to Egypt throughout its history often did not mention the mode of transport – at least not until the nineteenth century – and would simply refer to travelling or sailing north. Regardless, as a writer of social history, the fact that any records are left about any aspect of these journeys is a wonderful thing, and they will be used extensively throughout the following chapters to uncover the personal stories of these ancient hodophiles (those who love to travel).

## Written Records

To tell these stories I will be drawing on personal diaries, written records, religious texts, graffiti at sites, letters, stelae, statues, guidebooks, blogs and newspaper reports; all written records which help to tell the story of 5,000 years of travel. Many felt compelled to record what may have been monumental visits in their lives, right up to the modern era, and we are honoured that we can once more share their joy.

Unfortunately, due to the nature of the records prior to the nineteenth century we will only be able to examine a small section of society – the upper-middle and upper classes – as not only was it more challenging for the peasant farmers and very poor to afford the time and expense of travelling, they were likely illiterate and therefore are invisible in the written and to a large extent the historical record.

Literacy in the Graeco-Roman period is thought to have been at approximately 25 per cent and in the pharaonic period this was as little as 1 per cent of the population. This is reflected in the quantity of the written records about travel from the Graeco-Roman period. However, this means that there could have been more people travelling and sightseeing than records suggest as they only represent between 1 per cent and 25 per cent of the population. As time progressed, and literacy increased, by the seventeenth and eighteenth centuries more records survived even though the social classes remained the same. It is only in the modern era that all classes have the opportunity to travel, and to record their experiences through books, blogs and videos on social media.

When studying 5,000 of history, including visitors from all over the ancient world, it is necessary to understand the various languages in which the tourists wrote and how this changed over the centuries. Egypt was not a homogeneous society and over 5,000 years, culture, religion and language changed a great deal.

- *Hieroglyphics* (2686 BCE–391 CE) – This was the oldest writing system in Egypt and was seen in temples and tombs comprising pictures in place of letters and was used from the pre-dynastic period. As a language, it is phonetic with each sign representing both a concept as well as a sound (for example a cat, can mean 'cat' or the phonetic miw). It was generally only used by priests and therefore in 391 CE when the temples were closed and pagan cults shut down in favour of Christianity, the language was lost as the priests died. It was not until 1822 when Jean-François Champollion deciphered the script using the Rosetta Stone that the hieroglyphs could speak once more.

- *Hieratic* (3200–700 BCE approx.) – This was the shorthand version of hieroglyphs which could be written quickly and easily with ink and reed pen. It was essentially the Egyptian language written in the Greek alphabet. This was used for most non-religious or ceremonial writing, from accounts to stories to letters to graffiti.
- *Demotic* (700 BCE–452 CE) – This was a derivative of the hieratic language used from the twenty-sixth dynasty onwards, with the latest Demotic text dating to 452 CE from Philae. It is generally thought to be a written language only and was referred to by Herodotus as 'the script of the people'.
- *Greek* (332 BCE–30 BCE) – Following the invasion of Alexander the Great (332–323 BCE), the Ptolemaic dynasty was primarily Greek speaking, meaning this was the lingua franca for the military, government, education and religion. This started in Alexandria and moved via the elite throughout Egypt, but the ordinary Egyptians continued speaking their native Egyptian.
- *Latin* (30 BCE–641 CE) – Latin was the language spoken by the Romans and, therefore, Roman visitors left inscriptions in their native language. It was never a language used by the Egyptian people. The Romans annexed Egypt in 30 BCE at the end of the reign of Cleopatra VII (51–30 BCE), meaning Latin was more common after this period.
- *Coptic* (second century CE onwards) – This was the language of the Coptic Christians who retained the spoken, ancient Egyptian language, but wrote using a combination of Greek and Demotic letters. Church services in Egypt have elements in Coptic to the modern day.
- *Arabic* (640 CE–Present) – After 640 CE, Arabic became the national language of Egypt which slowly replaced Coptic and is still the language used today. Modern Egyptian Arabic is different from the Arabic used in other Arab states, but due to the export of Egyptian movies, the Egyptian dialect is understood throughout the Arab-speaking world.

Through the examination of the literary record we will also look at the cultural associations of travelling, and how it was considered within the villages throughout the pharaonic period. For example, the idea of long adventures was often the remit of fairy stories and appear in a number of literary tales from the Middle Kingdom, such as the *Shipwrecked Sailor*, the *Eloquent Peasant* and the later Demotic story of *Setne*. The concept of travel was not unknown, but these tales were aimed at a community which primarily stayed within their villages.

## How to Use This Book

The book is arranged in reverse chronological order, starting with the modern era prior to the 2011 political revolution and then moving back in time to the pharaonic period. Additionally some chapters have been themed to avoid repetition across the ages; for example, pilgrimages (chapter 4) and visiting oracles (chapter 7) were continuous practices (albeit changing ones) for thousands of years and it is more beneficial to discuss these as individual topics than spread over numerous chapters.

You will also notice as the book progresses that the term 'tourist' and 'tourism' becomes a lot looser in meaning as these modern terms can only really be applied to the eighteenth to twenty-first centuries in their truest meanings, and to a certain extent during the Roman period. In the later chapters they are taken to mean travelling for a purpose: research, education, pilgrimage, business or trade.

One thing that I hope is apparent as you travel back in time, is that the changes in tourism as an industry are clear. From the Roman period onwards it was realised that there was a lot of money in travel and, therefore, an important income for individuals and the country. This led to a tourist industry, the basics of which are still in place today.

Following the influx of tourists in the nineteenth century and the introduction of the package tours (chapter 2), tourism has been an important aspect of Egypt's economy where at its height in 2010 it brought in 13 per cent of the annual GDP.[23] This dropped off substantially following the revolution in 2011 and then took a further hit in 2020 with the pandemic (epilogue). But this is not the end of the story.

Travel to Egypt has been happening for more than 5,000 years and will likely continue for another 5,000 years, but how that looks could be very different. What this book also will show is that the basic motivations for tourism have not really changed in all that time; people travel out of curiosity, for work, for religion and for intellectual interest. Another thing that has not changed is the desire for people to write about their experiences from Herodotus, who produced the earliest guidebook, to the graffiti left by Roman poets, from the diaries of Victorian travellers to the travel blogs of modern influencers. With 5,000 years of travel we also have 5,000 years of travellers telling others about their experiences.

The main objective with this book, in addition to showing another dimension of Egyptian history, is to show that although the ancient Egyptians may be separated from us by 5,000 years, they were the same: they enjoyed visiting monuments of historical significance, they enjoyed speaking with people from different cultures and they enjoyed trying new things. They may not have had the same conveniences which we are fortunate to have today, but they managed with what they had to explore and enjoy the world they lived in.

So buckle up and enjoy the ride.

# Chapter 1

# The Importance of Tourism in the Twentieth Century

Tourism was at its height in Egypt just prior to the revolution in 2011. It had been a continually evolving beast, based on the foundations laid out by the pioneers of tourism in the late nineteenth and early twentieth centuries. Many of these staples to travel set in the nineteenth century can almost be seen as a blueprint for the ideal holiday (chapter 2). A standard itinerary for many modern tourists include:

- A cruise down the Nile
- A camel/horse ride at the pyramids at Giza
- Visiting the Valley of the Kings (especially the tomb of Tutankhamun)
- Visiting Karnak Temple
- A felucca ride at sunset

Although formulated in the nineteenth century, the tourist experience has changed a great deal since then, with amateur archaeology ending in 1923, and climbing to the top of the pyramids being prohibited in the 1980s. However, new experiences have been added including sound and light shows, day trips from the Red Sea to Luxor or Cairo and the ancient Egyptian theme park, Dr Raghab's Pharaonic Village in Cairo which opened in the 1980s. This is not generally on the tourists' must-see agenda for their first trip to Egypt, but it may make the itinerary on the fourth or fifth.

The Pharaonic Village (FIG 2) is a boat ride through the past, with dioramas of ancient Egyptian temples, homes and workshops complete with actors in modestly appropriate costumes. The uniqueness and strangeness of the theme park is summarised by S. Slyomovics:

> The physical proximity of the authentic artifact to its replica is especially disconcerting to Americans, for although we are accustomed to a variety of simulacra representing the past reserved in full-scale copy, it must be a past removed not only historically but necessarily geographically too. ... While a Pharaonic Village erected in Disneyland would possess a certain

touristic logic, such a construction around the corner from the pyramids and the Sphinx does so less convincingly.[1]

Her point was that the Giza and Saqqara pyramid fields, as well as the Egyptian museum, are literally only a few kilometres away, so she felt it would be more appropriate to visit them than this pastiche theme park. Additionally, whilst Dr Raghab's is a bizarre and entertaining afternoon, it would be better placed in Sharm el Sheikh or Hurghada which have no ancient monuments and would probably be more of a draw for Western tourists.

## Tourism Revenues

The 2011 Arab Spring saw tourism revenue and tourist experiences plummet dramatically, and even once it had started to rebuild, the 2020 pandemic, which prevented all travel, once more brought the Egyptian tourism industry back to ground zero.

Prior to this prolonged period of upheaval, tourism was the largest contributor of foreign exchange earnings in Egypt.[2] In 2010 for example tourism accounted for 13 per cent of Egypt's GDP, 10.9 per cent of employment and generated $14 billion of export revenue, or 22 per cent of total exports.

2010 represented the pinnacle of the tourism industry and had been on a relatively stable upward trajectory since the nineteenth century. The nineteenth century introduced Thomas Cook cruises and the opening of the Suez Canal, which was important for international trade, as Egypt became a stopping point for people travelling east from Europe, whether on business or for leisure.

By the end of the twentieth century and the introduction of cheap air transportation, both trade and tourism started to grow exponentially and is reflected in the figures. Between 1982–83 tourism receipts amounted to $304 million and rose to $6.429 billion in 2004–5. At the beginning of the twenty-first century tourism receipts represented 20 per cent of the GDP,[3] which had risen from five per cent in 1980–1 and seven per cent in 1990–91.[4] Tourism was clearly big business.

The numbers of tourists arriving rose from 254 million in 1990 to 507 million in 2008.[5] Without the Arab Spring, this trajectory was likely to have continued to rise at the yearly average of 21.5 per cent until the Covid outbreak in 2020.

Tourism was not just an important contributor to Egypt's economy, but also to the local communities and between the 1970s and 1990s, approximately one million local people were employed directly in the tourism industry and four to five million indirectly.[6]

The increase in tourism was not just with travellers from Western countries as Egypt had become an attractive option for four main groups of travellers:

1. Tourists from west and south Europe – visiting the Red Sea.
2. Tourists from west and south Europe – visiting Cairo, Luxor and Aswan.
3. Tourists from Arab states (Gulf including Saudis, Kuwaitis and Palestinians) – visiting Cairo and Alexandria.
4. Rich Egyptians and ex-pats living in Egypt – visiting Alexandria and Suez Canal.[7]

## Terrorism

Under President Nasser (1954–1970), Egypt fast became a popular tourist destination, but this was unstable due to local political instabilities, terrorism and natural disasters.[8] These included the Arab-Israeli conflict, the assassination of President Sadat (1981), the Israeli invasion of Lebanon (1982), the hijacking of a cruise ship (1985) and the bombing of Libya by the US (1986).

Each of these events resulted in a dip in tourism numbers and a subsequent knock to the economy. The Gulf War in 1991, for example, is thought to have resulted in a 50 per cent drop in tourist revenue.[9] The global financial crash of 2008 obviously affected Egypt financially as it did the rest of the world, but the following global recession meant there was reduction in tourists travelling to Egypt from Europe and the Gulf States. This did not start to lift until the second half of 2009.[10]

However, the biggest impact on tourism prior to the 2011 revolution was terrorism with the *al-Gama'a al-Islamiyya* (The Islamic Grouping) targeting tourists in the 1990s. There were a number of incidents, including; the Luxor massacre in November 1997 where fifty-eight foreign nationals and four Egyptians were killed by six terrorists disguised as security guards at the temple of Deir el Bahri; a bomb exploding on Pyramids Road, Cairo near a tour bus killing two Egyptians and wounding twenty-two tourists in June 1993; gunmen firing at a cruise boat in 1994 in Upper Egypt killing a German tourist; and in September 1997 a gunman shooting eleven German tourists in the car park of the Egyptian Museum in Cairo. These are just a few incidents of more than twenty terrorist attacks aimed at tourists in the 1990s, which clearly affected the visitor numbers to the country. Between 1992 and 1995 lost tourism revenue was thought to be in the region of US$1 billion.[11]

After the Deir el Bahri massacre in 1997, the airports closed for several months meaning no tourists could enter the country and even after they were opened it took a number of months before tourists were confident enough to return.

Even though on the whole Islamist groups did not operate in the Luxor region, and 'religious leaders, popular sentiment, and a pervasive secret police are strong enough to keep them out',[12] tourists still chose to stay away on the grounds of safety, many at the advice of their own governments. The Egyptian government kept announcing they had 'successfully eliminated all militants, only to be proven wrong when yet another militant attack took place'.[13]

These attacks on tourists and, therefore on Egypt's GDP, ended up becoming political with President Mubarak claiming Egypt was under an attack of destabilisation from Islamic movements such as the Muslim Brotherhood, al-Qaeda and Hezbollah which threatened everything Egypt stood for and prevented a move towards liberal democracy. As E. Wynne-Hughes refers to it, it was viewed as 'bad' Muslims against 'good' Muslims.[14]

There was a further major decline in tourism in 2001 and 2002 following the 9/11 attacks and the subsequent Western war on terror in Afghanistan and Iraq. However, by 2003 the figures once more had risen indicating that most dips in tourism at this time lasted only for a season or two.

This quick rise could be attributed to the change in perceived danger following 9/11 where all Western countries and Westerners were potentially at risk from Islamic fundamentalist groups within their native countries. Therefore, they were as likely to be a target in their local supermarket or high street as they were in Egypt.

## Damaging Monuments

Although bringing a lot of money into the country, the sheer numbers of tourists travelling to Egypt since the nineteenth century has been problematic. In the summer months in Alexandria for example there can be as many as 2.2 million tourists from Egypt and other Arab countries. As the population of Alexandria stands at only four million people an increase of 50 per cent of people can tax roads, traffic, utilities and services to breaking point, and they therefore have a detrimental effect on local communities.[15] It goes without saying that having this many additional people descending on towns and cities needs to be managed, and before 1998, 25 per cent of foreign investment went towards the tourist industry in order to develop more than 800 establishments for tourists such as hotels, restaurants and visitor centres.[16]

The number of tourists at the ancient sites were (and are) also having a detrimental effect on the monuments themselves, leaving many damaged beyond repair. The damage varies from that caused by vibrations of thousands of footsteps every day, to a build-up of condensation in enclosed spaces like tombs, from tourists' sweat and breath which results in plaster and paintwork crumbling.

The humidity level in a tomb for example, can rise from five per cent at dawn to 90 per cent at the end of the day with temperatures rising to 40°C. Without tourism humidity remains stable at about 40 per cent and temperature at about 25°. Such drastic changes can cause a build-up of sodium chloride (salt) which can cause the paint and plaster to fall off the wall. At Abu Simbel, in the far south of Egypt, at the high point of tourism there were as many as 2,000 tourists at the site every 90 minutes. The vibrations from their feet, their voices and their transportation caused irreparable damage to structure.

There is also additional damage carried out through vandalism (chipping a piece of pyramid off to take home as a souvenir), graffiti, and touching, sitting on or leaning against the monuments which damages paintwork and carvings, as do scratches caused by bag buckles and spokes from umbrellas used to avoid the harsh Egyptian sun.

The removal of artefacts and structural blocks has been a problem since the Graeco-Roman period (see chapter 6), and throughout 5,000 years of tourism, graffiti and touching the monuments has remained commonplace, but the problem currently is the sheer numbers of people doing these things every single day. What makes this wilful and unintentional damage so frustrating is that any visitor to the Valley of the Kings, for example, will see a busload of people enter a tomb, march in and out again in a few minutes, without looking at where they are and absorbing the artwork and history their very presence is destroying. All so they can tick 'Valley of the Kings' from their check list.

Although outside the scope of this book, environmental and social issues are also adding to the damage to the monuments. For example there was a great deal of damage done to the Valley of the Kings following the flash floods in the autumn of 1994. Tutankhamun was one of only four tombs not penetrated by the water. Additionally, the lower parts of the walls at Karnak temple are being damaged by the drainage from the agricultural land which runs off towards the Nile going through the temple itself. This has resulted in the walls being so encrusted with salts that the decoration has been obliterated and images are no longer visable.

Similar concerns were also apparent in Alexandria when in 1997 more than a third of all Alexandria's wastewater and untreated sewage was pumped directly into the Eastern Harbour, and the remaining two-thirds was pumped into Lake Mariut to the south. Whilst this in itself is a disgusting means of disposing of waste, it affects the archaeological remains which are under the water in the harbour.

The harbour preserves the remains of the Pharos Lighthouse, the Ptolemaic royal quarters and at least three French wrecks from the 1798 Battle of the Nile, which took place between Napoleon Bonaparte and Lord Nelson; *L'Orient*, *Le Guerrier* and the frigate *L'Artémise*.[17] Although these sites are valuable to archaeologists, due to their general inaccessibility they have not been a draw for tourists and their welfare, therefore, has been greatly neglected. In 1882, Henry Gorringe commented that on a clear day in the Eastern Harbour it was possible to see the archaeological remains beneath the water, but by the end of the twentieth century, due to the raw sewage, visibility was greatly reduced.

This problem was addressed at the International Workshop on Submarine Archaeology and Coastal Management (SARCOM) in 1997, which was attended by an interdisciplinary team, UNESCO and the Ministry of Culture.[18] By 2003 (some six years later) three main sewerage outlets into the archaeological area of the Eastern Harbour were closed with a stipulation that they were only to be opened after heavy storms.[19] Subsequent to these closures, four diving companies were opened in 2001 allowing tourist dives over the Pharos area, although there was still no access to the harbour due to pollution. In 2006, when I went diving here, the visibility at the Pharos site was still so poor that even under the water it was impossible to make out much of the archaeological remains.

If raw sewage into the Eastern Harbour was not damaging enough, in 1993 poor research into the area saw 180 concrete blocks weighing 7–20 tons each, being dumped into the sea near Qaitbay Citadel in order to create a breakwater. These were dropped directly over the remains of the Pharos Lighthouse beneath the water, destroying it.

This led to the rescue mission (SARCOM Workshop) in 1997, where more than 2,000 artefacts were excavated from the sea and put on display at Kom el Dikka (FIG 3), with others being sent to Berlin. Ten of the destructive concrete blocks were removed in 1996, forty-five in 1998 and the remainder of them January 2001. This work meant the site was now ready for archaeological exploration.[20]

The improvement of Alexandria was an ongoing project, and the 2005 Comprehensive Master Plan for Alexandria was intended to 'preserve the unique coastal character of Alexandria as an important tourist centre'. It was to focus on the historical heritage whilst also improving the infrastructure of the city including clean water, roads linking the east and west of the city, more green spaces and waterfront places for urban development,[21] although whether this was to improve the city for the locals or the 2.2 million tourists a year was not made clear.

Alexandria has always posed a bit of a conundrum as the majority of the ancient monuments are under the sea, meaning they are difficult to access and therefore difficult to utilise to attract high-spending Western tourists. The SARCOM workshop in Alexandria in 1997 discussed how to make the underwater artefacts from the Eastern Harbour more accessible to tourists and came up with three rather startling ideas:

1. Drain the harbour exposing the remains. Rejected.
2. Create a new two-level museum – one at shore level displaying artefacts already lifted from the seabed, and the other a floating museum on a U-boat or a glass bottomed boat to view the underwater archaeology.[22] The boat aspect was rejected.
3. Create a three-level museum – one level as an onshore building with artefacts from Alexandria as a whole, the second floor showing artefacts from the seabed exhibited in aquariums and, the third, an underwater tunnel going through the Ptolemaic city.[23] However, the Plexiglas tunnel was thought to be limiting, allowing visitors to see only one archaeological zone.[24]

At the time of writing none of these ideas had been commissioned despite grand ambitions for their success at the time: 'Once the underwater archaeological museum has been developed and the area has been cleared of polluting sediments and the city's wastewater is properly treated, Alexandria will become a centre of cultural tourism.'[25]

There were a lot of elements to correct before any of these ideas could be started. However, it does need to be considered that if any of these plans had been accepted there would be a danger of losing some aspect of the natural and cultural environment: 'The fact that matters, however, is that once a hotel or a resort is built near the seashore, for example, the natural habitat for indigenous species is damaged.'[26]

It could also be argued that Alexandria itself already has several ancient monuments which could be marketed and promoted to draw tourists, such as the ancient town at Kom el Dikka, the catacombs of Kom el Shuqafa, Pompey's Pillar, the remains of the Serapeum, Qaitbay Citadel, as well as the museums. Rather than creating experimental and potentially damaging resorts, why not encourage tourists to visit what is already preserved?

## Increasing Tourism

No matter how popular Egypt was with travellers, the Egyptian government have been striving to increase numbers and the revenue they generate since the 1960s. Initially, the growth of tourism as well as the improvements to infrastructure were slow, creating a chicken and egg situation. There had to be reasons for tourists to visit, places for them to stay and things for them to do, but to provide these there needed to be financial investment.

In 1958 the Nile Hilton in Cairo opened and became a major draw for international tourists. Although the land was still publicly owned, the hotel itself was run by private investors.[27] This was followed in 1961 by the introduction of the Sound and Light Show at the Pyramids at Giza, which is still running sixty years later. It is now overseen by the Misr Company for Sound and Light, alongside similar shows at Karnak (1972), Philae (1985), Abu Simbel (2000) and Edfu (2010). Additionally, in 1972 and 1973, a special exchange rate was introduced, and duty-free shops accepted currency other than the local Egyptian pound to make it easier on the pockets of tourists and, therefore, a more attractive destination.[28]

However, the Egyptian government realised more was needed, and to make Egypt a more attractive place for investors in the 1960s and 1970s, various tax advantages for tourism companies, such as hoteliers and other tourist facilities, were introduced. This included up to twenty years of tax exemption and concessions for imports and tariffs.[29]

As tourism increased, Egypt no longer only appealed to middle class Europeans, Japanese and Americans, but also by the 1990s, to the lower middle classes which meant marketing and finding new ways to attract tourists had to be accelerated as the customer base had increased.[30] In 1998 the minister for tourism, Mandouh El-Beltagui said:

> lodging capacity was raised from 18,000 in 1982 to 75,000 in 1997. And we now have 603 projects under construction ... Egypt has invested more than $50 billion ... to improve the infrastructure and public utilities.[31]

There were a lot of changes from the 1990s through to the early noughties and the authorities were treading a very fine line between improving tourist experiences and facilities and preserving the monuments, as well as the indigenous natural environment and the indigenous culture. It is only in recent years that this has been addressed more evenly.

Some of the greatest changes since the 1990s have unsurprisingly been in the areas with the most tourist footfall, namely Luxor which houses Karnak

Temple, Luxor Temple, Deir el Bahri, the Valley of the Kings and Deir el Medina just to name a few of the sites. Zahi Hawass, the Former Minister of State for Antiquities Affairs of Egypt, commented in 1998 that: 'Luxor is a unique city. Its atmosphere of quiet, ancient streets, and culturally special people should be preserved.'[32]

However, in direct contrast to this statement, 2008 saw the remodelling of the East Bank Corniche at Luxor. This scheme aimed to turn Luxor into a tourist centre, with shopping malls, restaurants and plenty of parking for coaches and taxis without really considering that Egyptians live there. With more than 250 coaches a day coming to Luxor from the Red Sea for day trips, parking and traffic was a problem, and the plan included widening the Corniche, creating coach parks with greater capacity and providing a walkway alongside the river, destroying any remnants of the 'quiet, ancient streets'.

In order to create this touristic centre, many buildings along the Corniche were demolished which included nineteenth-century colonial buildings, with Chicago House, home to the Epigraphic Survey of the Oriental Institute since 1924, losing half its garden and the Mercure Etap hotel losing its front wing.

## Enclave Tourism

The Corniche plan was seen as controversial as it very much separated the tourist experience from the local Egyptians, creating segregated areas and no-go areas for the locals. Add to this, the increased space for coaches and cars that would attract more visitors to Luxor further exacerbated the problems the changes were meant to alleviate. This was the start of enclave tourism which completely separated the tourists from the local Egyptians and is something that continues to this day. Modern buildings like hotels, malls, restaurants are: 'quasi fortresses, which are largely inaccessible to most urban residents except as workers'.[33]

This was clear to any tourist wishing to thank their tour guide or taxi driver with drinks in the hotel. The locals were stopped by the security on the door and forbidden entry. The authorities even went so far as to close monuments like the pyramids at Giza on public holidays as a means of preventing locals from visiting.[34] Fortunately since 2011, this policy seems to have stopped as many of the monuments have more Egyptian than foreign visitors and there is a local price and a tourist price to make it affordable for local residents, but this does not mean local Egyptians are welcome in hotels and restaurants which cater solely for tourists.

So, what led to this segregation? The World Bank conducted a survey in the 1980s in Luxor on behalf of the US consulting firm Arthur D. Little and discovered that the tourists' biggest complaint was the hassle from the locals to enter their shops, go on their feluccas or jump into their kalashes, and prior to the revolution in 2011, a common tourist complaint was that it was impossible to walk along the street without being followed and harassed. More so as a woman.

The World Bank's programme was to increase tourism and, therefore, on the back of their survey, pedlar licenses were no longer issued and a visitor plan was put in place to separate the tourists from the locals, which even included ideas such as covered walkways from the coach parks to the temples bypassing the villages.[35] Another part of the 1980s development programme was to build a new embankment along the Nile at Luxor. However, rather than bringing employment to the local people the work was outsourced to China. Between 1982 and 1992 tourism to Egypt almost doubled, bringing in a great amount of revenue which unfortunately went to large international hotels and cruise ships rather than the local economy, even though they employed local people.

With such an increase in tourist numbers it was important to improve visitor management, further increase the flow of tourists and improve local infrastructure (water, electric), as well as attract wealthier tourists by limiting temple and tomb capacity.[36] This capacity limitation is still being used today with the tomb of Nefertari in the Valley of the Queens costing about £50 to enter, with a very limited number of tickets per day. They always sell out.

Whilst the tourist walkways never came to fruition, the idea of the tourists and locals being segregated is presented in all marketing and guidebooks, where the ancient monuments are used as the draw for tourists with images only showing Western tourists at the sites. Images of local people are limited to quaint experiences like the village visits, or the desert camp outs where they are shown in traditional galabeyas carrying out traditional tasks like pottery, baking bread or using a traditional shaqqia.

Ironically, whilst on one hand the authorities advocate enclave tourism by separating the local population from the tourists as much as possible, on the other, they create 'fake interaction with locals through set visits to alabaster factories, papyrus factories (often mislabelled as museums), perfume factories or carpet factories. These manufactured encounters are meant to represent an experience where the visitor is provided with refreshments, a demonstration of the craft and then left to browse – sometimes for longer than they spent at the ancient monuments. The tour guide gets commission on every sale, so it is a lucrative endeavour. Some tourists love it, others see it as an inevitable low point of a guided tour.

## Thomas Cook's Legacy

Another issue which is particular to Luxor and has been consistent since Thomas Cook introduced his first cruise liner (see chapter 2) in 1869 is the number of cruise boats lining the banks of the Nile. Prior to 2011 these could be four deep on the east bank of the Nile at times, marring views but also pumping oil directly into the Nile and diesel fumes into the atmosphere. Cruise boat numbers have been growing exponentially since the 1980s:

- 1982 – 47 cruise boats
- 1989 – 119 cruise boats
- 1998 – 200 cruise boats
- 2000 – 276 cruise boats[37]

In 1989 the government introduced a temporary freeze on cruise boat licences to reduce the numbers but as soon as it was lifted, the numbers started to rise again. The Egyptian government seemed to be unable to find the balance between attracting more tourists and reducing the problems of cruise boats, coaches and overcrowding.

In 2008, the Egyptian authorities were still trying to deal with the cruise boat issue and proposed a new cruise ship port further south of the Nile, past the Luxor Bridge, with the intention of removing the boats from central Luxor. This was a clear example of moving the problem from the 'Instagram worthy' part of central Luxor and to an area where it would only affect native Egyptians. At the time of writing this, a port has not been commissioned and cruise boats can still dock anywhere along the east bank at Luxor.

## Tourist Management

With tourism forming such a large part of Egypt's GDP, deciding how to manage tourists was always a delicate balance; on one hand it was important to maintain the economy and attract more tourists, but on the other hand it was also essential to maintain the natural environment and conserve the monuments whilst also dealing with social requirements and political restraints.[38]

Even with the continual increase of tourists to Egypt, and the associated problems by the end of the twentieth century, the country was competing with other international holiday destinations, meaning officials felt it was necessary to expand the tourism industry quickly in order to remain relevant within a global industry.[39] It was these conflicts of interest which meant that the balance between the economy, the tourist experience, the environment, the safety of the monuments and the welfare of the local population has not always been handled well.

As Zahi Hawass said in 2004:

> The economic need to accommodate more visitors and to encourage them to stay longer has to be balanced with the long-term future of the monuments themselves. More careful regulation of visitors and traffic (both vehicular and animal) combined with judicial opening of more areas of interest is vital to achieve this end. To implement and maintain this, it is essential to establish local training centres for conservation, restoration and scientific archaeological work.[40]

So, at that time, it was thought that opening more monuments would essentially thin out the crowds at any one monument. In theory it could work, but coupled with the other plans to attract more tourists, it could have simply resulted in the same numbers of tourists at more sites. Realistically, it needs to be considered that no new 'unknown' monument is going to be chosen over the pyramids of Giza or the Valley of the Kings for a tour group on a day trip from the Red Sea or someone visiting Egypt for the first time.

There is no avoiding that it is the ancient culture that attracts visitors to Egypt, and therefore the ancient past has become an important aspect of modern Egyptian identity. Consequently, restoring the ancient monuments and creating outdoor museums is important as a means of presenting a nationalistic front to the world, as well as connecting Egypt with its glorious ancient past.[41] However, like the Golden Age of Travel (chapter 2), this glorious past does not really exist and is a marketing tool where the best of the past is chosen to represent the continuity of the present – and inevitably this means the over-saturated sites of Giza and the Valley of the Kings. Trying to market Deir el Medina or Coptic Cairo as a draw for tourists is not going to have the impact of the pyramids, or Tutankhamun's tomb.

## Replica Monuments

An alternative plan to opening more monuments was to build replica ancient tombs as a means of encouraging visitors away from the original to the 'copy'. This was considered quite controversial regarding the tourist experience, but also in regard to building what are essentially concrete buildings in historical areas.

By 2014 two tombs had been reconstructed: that of Thutmosis III and Tutankhamun. The reconstructed burial chamber of the tomb of Thutmosis III went on a worldwide tour and stopped in Edinburgh in 2015 under the exhibition name of *Immortal Pharaoh: The Tomb of Thutmoses III*. The burial chamber was recreated by a Spanish company called Factum Arte (Factum Foundation) for the National Archaeological Museum in Madrid. At the end of 2018, the burial chamber had been installed in Bolton Museum as part of its new Egyptian galleries and will be a permanent feature. Whilst an interesting approach, and a great feat of engineering, will this prevent people from wanting to view the original? Or encourage them to go and see the rest of the tomb?

Having reconstructions of the tombs outside Egypt will not necessarily deter people from visiting Egypt and the real thing, which is why the reconstructed tomb of Tutankhamun was created as a static structure near Carter House on the west bank at Luxor. It opened in 2014 and was considered an innovative approach to tourist management. The idea behind this replica was to draw the crowds away from the original in the Valley of the Kings whilst enabling them to still experience the tomb.

This replica tomb was also created by Factum Arte (Factum Foundation) who laser scanned the original tomb images and printed them onto a fibreglass mould of the tomb. It was finished with a fine, coloured, gossamer skin attached using a vacuum process. The reconstruction also included the decorated wall fragment that was removed by Carter's team during the original excavation to enter the burial chamber. The original has been lost since the tomb was discovered and was recreated from Harry Burton's photographs. This is displayed in the annexe room along with information on the reconstruction and is something that makes the replica an 'improvement' on the original. Adam Lowe from Factum Arte believes the replica to be accurate 'to around a tenth of a mm'.

Factum Foundation had also been working on the tomb or Sety I for seventeen years, with the Antikenmuseum Basel, the University of Basel and the Ministry of Antiquities in Egypt. In 2017 facsimiles of the Hall of Beauties were put on display along with a facsimile of the sarcophagus and tomb fragments in museums across the world in an exhibition called *Scanning Seti: the Regeneration of a Pharaonic Tomb*. Once the exhibition finished, the rooms were to be donated to Egypt with the intention of being installed next to the replica tomb of Tutankhamun.[42] At the time of writing this had not yet happened.

The controversy comes in with the discussion as to whether a replica, no matter how detailed and how accurate, is the same as the real thing. For many visitors to Egypt, who are ticking sites from their bucket lists without really looking, this should not make much of a difference, but would they choose a replica if the original is open? Probably not.

I was lucky enough to be working in Luxor when the reconstruction of the tomb of Tutankhamun was opened and I was invited to the opening ceremony. And whilst you cannot fault the visual elements of the reconstruction as they are perfect, the 'tomb' itself felt new, the lighting was too bright, and had the atmosphere of a modern building. There were a lot of very hot and bothered people in there so the temperature and smell were similar, but it was not the original. It was very confusing.

Even the creator, Adam Lowe, knows the acoustics are not right and hopes to get engineers in to recreate the acoustic signature of the original tomb.[43] But would a first-time tourist know the difference? Probably not, and it would protect the original, but only if the original was unavailable to visit.

This conundrum still has not been resolved, even following the 2011 revolution and the 2020 pandemic. One thing that is certain however, is that if both tourist and environmental issues are not addressed, the monuments of Egypt which have stood for thousands of years, could only survive for another 200 years, although probably less. The problem seems to be the conflict of interest between the preservation of monuments and attracting more tourists, and it has been addressed in various site management plans for the past three decades. One of the main criticisms of most of these plans is that they favour the foreign (Western) tourist over the needs of the local people.

A particularly controversial plan was the destruction of the village of El Gourna, which was situated above the Valley of the Nobles in Luxor. For tourists to reach the tomb entrances they had to walk through the village, which had been on the site for more than 150 years.

Some of the village houses were considered historical monuments within their own right as they included murals associated with the hajj, which is when dedicated Muslims make the pilgrimage to Mecca. However, despite the village's historical past, over the past century and a half, the village location has caused irreparable damage to the ancient tombs beneath including theft, water damage and vibrations from day-to-day living.

In the early 2000s the government earmarked the village for demolition, evacuating the 30,000 villagers with promises of better living conditions and utilities. The site was to be excavated, exposing all the tombs in the area and making it the biggest archaeological site in the world. Many of the villagers were moved to inadequate, dilapidated apartments eight kilometres away, breaking up the close-knit community, as well as forcing some large families of ten or twelve into two-bedroomed flats.[44]

This scheme formed part of a country-wide plan which included moving people living and working at the pyramids in Giza, Islamic Cairo and other historical sites in Luxor. Often the government justified their means of removal by presenting the local villagers as criminals who had been looting the sites and who were generally uncivilised, and living in the wrong place.[45]

## Controversy of Camel Rides

Other changes concerned the general tourist experience, and one controversial element was the issue with animal rides. Anyone who travelled to the pyramids at Giza in the second half of the twentieth century and the first ten years of the twenty-first century could not help but notice the increase in animals there: camels and horses by the dozen, with aggressive and persistent sales techniques to persuade tourists to experience the pyramids from the back of an animal. They did not always take no for an answer and left many tourists feeling harassed and unable to enjoy the monument which was the reason they were there in the first place.

However, there had been long debates about whether animals should be allowed within the boundaries of the pyramids at all. In 1998, Zahi Hawass, then Director of the Giza Plateau, commented that camels and horses polluted the site, damaged the monuments and bothered the tourists. Therefore, the Egyptian authorities put in legislation to ban animals from the site, restricting them to south of the third pyramid. There were further concerns about the welfare of the animals as they worked long hours in the desert heat which in the summer can reach over 50 degrees with little shade or water and horses would often slip on the tarmacked roads. The animals were generally worked until they collapsed with exhaustion and their owners often beat them until they got back up. However, despite this, the 1998 legislation was not enforced and made very little difference to the number of camels and horses at the site.

In 2007, Zahi Hawass, at this point Secretary General of the Supreme Council of Antiquities, stated in an interview that, 'The people here have been handed a gold platter – the pyramids ... Instead of guarding it, they (defecate) in it.'[46] He once more vowed to ban animals, bogus guides, wandering vendors and anyone else who is looking to make money unofficially from the tourists.

However, with bribes to the police as an acknowledged payment to get through the wall surrounding the site, there were still camels and horses on the Giza plateau until 2020 despite protestations from the government that they were no longer there.

In fact, it was not until 2020 that Egypt's Ministry of Tourism and Antiquities once more announced 'plans' to ban the animal rides at the site.[47]

Only time will tell if these plans are ever enforced, and with camels featuring in many of the promotional shots of the pyramids, there are some very mixed messages about what tourists are to expect when they visit this site.

## Tourist Not-Spots

Many of site management plans and investments were focussed on tourist sites which were already at, or over, capacity, as a way of managing these monuments rather than drawing attention away from them. Few plans addressed the repair or conservation of sites which were equally historical but not visited by tourists.

For example, there are more than 300 Islamic monuments in the old quarters of Cairo with 47 per cent of them in a dilapidated state, closed to the public, used for drugs and prostitution and generally neglected. Those which do attract tourists and visitors like Sultan Hassan Mosque, the Citadel and Al-Azhar Mosque are, conversely, kept in a beautiful condition and are well-funded.[48] This approach can be seen in many areas of Egypt as soon as you go off the beaten track – such as the mastaba graveyard at Giza which, even though only two minutes from the pyramids, remains mostly unvisited and tomb entrances are full of rubbish. Many of them have been used as latrines.

It is a tricky situation for a weak economy like Egypt's as it has thousands of historical monuments which all need conservation, reconstruction and protection but there are limited funds in which to do this. Therefore, it is not surprising that money is diverted to those monuments which draw tourists and bring money into the state and the local economies. On the other hand, according to many of the government statements, they are trying to spread tourists out in order to reduce overcrowding and they have more than enough monuments to do this without building replicas of tombs and temples.

Unfortunately, this leads to another important aspect of tourism which is ease. Many tour guides want to take their tourists to easily accessible monuments, to speed up trips so as to allow more time spent in the alabaster and papyrus factories which is where they make their money with commission. At high season in the 1990s, this meant there could be 1,500–2,000 tourists queuing to get into the tombs of Ramses IV, VI and IX in the Valley of the Kings – all near the entrance – when there were in fact twelve tombs open further into the Valley.[49] For example, the most remote tomb, that of Thutmosis III, which requires climbing a metal staircase high into the cliff face has never had 100s of people waiting to enter as it takes time and effort to get to it. Alternatively, the tomb of Ay in the Western Valley which is a short car ride away is also nearly always empty and is on tour itineraries as a specialised trip. This tomb is almost identical to the tomb of Tutankhamun, but bigger and is the only royal tomb with the fishing and fowling scenes normally seen in the tombs of the nobles. It is a beautiful and unusual tomb, but the guides just do not want to go that far into the Valley and cut into the shopping time.

## Successful Projects

Not everything that the Egyptian authorities have suggested, however, is detrimental to the monuments, or the local people, and in 2018 there was a massive breakthrough for tourist accessibility when Egypt's Minister of Antiquities, Khaled al-Anani, announced that there would be ramps at Karnak temple to make access easier for those with physical disabilities.[50] This was the first (and only) site in Egypt to introduce access ramps. Such small changes can make Egypt more attractive to tourists, especially the older generation who are often higher spenders than the young budget travellers.

Another long-term project that has almost come to fruition is the Grand Egyptian Museum, which is to replace the museum in Ismailiya Square (Midan Tahrir) which was built in 1902. This museum took five years to build, and it opened in 1902 to 500 specially invited guests. It has remained a must-see for foreign tourists for more than a century. However, this museum was found to be no longer fit for purpose, with more items in store than on display and hot and dusty rooms which were more damaging to the objects themselves – many of which could be seen crumbling in cases, even though they were wonderful examples of Victorian museum displays. More importantly, from a touristic point of view, there was a lot of traffic near the museum and, as it was downtown, tourists struggled to navigate the traffic when trying to cross the road.

The idea for the Grand Egyptian Museum was to make it a centre of learning for world scholars with a museum as well as 'a fully computerised information centre for Egyptologists', 'a publication centre for books, CDs, videotapes etc.', as well as 'proper laboratories for scientific research, conservation, restoration and photography according to the latest technology' and, of course, extensive restaurants and shopping facilities.[51]

The winning design for the Grand Egyptian Museum was presented by an Irish firm, Heneghan Peng Architects.[52] The museum was initially started in 2002, when the then President Hosni Mubarak laid the foundation stone, although the idea had been announced a decade earlier in 1992. Whilst some work started in 2005, the main construction work did not start until 2012, some twenty years after the idea was introduced. It was meant to open to the public in November 2022, although at the time of writing this had been changed to '2023'. It is set to be the world's largest museum dedicated to one civilisation, and hopefully will bring a lot of items out of storage and in inadequate display to the forefront.

This project has not been free of criticism, with comments such as it was 'unjustified to have a concrete building that is almost fully air conditioned' in a city which has regular power cuts, without any reference to the potential environmental impact. Further criticisms have been on the location, indicating it could have been built elsewhere in the city, in order to rejuvenate poverty-stricken areas, rather than in Giza.[53] Cynics could suggest that placing this museum in Giza makes the tourism journey easier as buses can stay in Giza to visit the museum, pyramids and a perfume factory without having to go through traffic and, therefore, maximise the day. It will also potentially make policing, security and tourist management easier and more effective.[54]

## Throwbacks to the Golden Age

In the face of such increasing tourist numbers as Egypt saw between the 1980s and 2011, travel, for some tourists, was a check-list exercise, a conveyer belt of tourist buses, whistle-stop tours of the most notable places ending with a visit to an overpriced factory or shop where the tour-guide gets a hefty commission. This unsurprisingly has led to some tourists seeking a different experience. There are many options for alternative adventures in Egypt but one of the first, and most popular, is the rise in 'Golden Age' travel opportunities.

This idea of a travel Golden Age can only really appeal to European and American travellers, as this period was one of colonialism and wealth. It is a troublesome ideology to recreate as it avoids modern Egypt including contemporary Islam and extreme urbanisation. For example, publicity shots of the pyramids and Sphinx at Giza show a wonderful desert landscape with sand for miles. However, in reality, simply turning the camera around will show the pyramids and Sphinx are in an urban area and it is possible to gaze upon the antiquity of the Sphinx from the comfort of a fast-food restaurant.

This reality is not what many tourists want, and therefore Golden Age travel is about selling the dream of becoming an explorer like Belzoni or Carter or even Indiana Jones, and the Lonely Planet claims 'Egypt brings out the explorer in all of us'.

Despite the troubling colonial overtones, the Egyptians have embraced this aspect of tourism, and one the opportunities for nostalgia of bygone times is being able to travel on an original paddle steamer, the *Sudan*, which is the only surviving vessel from the Cook and Son (chapter 2) fleet to still be operational.

The *Sudan* was launched in February 1921 and was in service until the start of war in 1939 when she was used as a floating officer's club. In 2003, she was sold to the Voyageurs du Monde, a French travel company who restored her enabling her to sail between Luxor and Aswan ten months of the year. Since 2013 cruise boats have once more been permitted to travel the full length of the Nile from Cairo to Aswan, and the *Sudan* has taken advantage of this.

Other traditional vessels on the Nile include the *Misr*, a twin-screw steamer (with paddles on the side purely for decoration) which was commissioned in 1918 and later sold to a tourist company in the 1990s, and the *Karim* which was built in 1917 which was also purchased by a tourist company in the 1990s (FIG 4). Thomas Cook's *Delta* is a floating restaurant in Cairo and the *MV Andrea* restaurant in Roda, Cairo was once the *Mahasen*.

In the 1990s the Voyageurs du Monde Company also started to revive the dahebeya as an alternative means of travelling in Egypt. Dahebeyas were traditionally used by tourists in the nineteenth century as houseboats, and accommodated servants, full dinner services, a library and even a piano (see chapter 2). In the 1990s a fleet was newly built according to original plans and photographs and in 2021 there were about forty on the Nile, and others used as houseboats for local residents. Egyptologist Kent Weeks followed in the footsteps of previous nineteenth-century archaeologists and lived on a dahebeya near the Mummification Museum in Luxor between 2003 and 2013.[55]

In the same way that the nineteenth-century travellers wanted an 'exotic' holiday where they could experience new things, so do twentieth and twenty-first century travellers. This has led to a number of 'adventure' holidays which move away from the traditional cruise down the Nile and nostalgia holidays and can include cultural tours off the beaten track, trekking in the mountains, diving in the Red Sea, camping in the desert or birdwatching. Some visitors to Egypt on these types of trips do not visit the ancient monuments at all as this is not their reason for being there.

Such alternative trips are not a new concept and in the 1980s the Egyptian authorities started promoting their beaches at Hurghada, Sharm el Sheikh, Dahab and Naama. Such beach and diving holidays grew in popularity and remain popular to the modern day. Prior to the 1980s the Red Sea and the Sinai Peninsula were closed due to the Egyptian-Israeli conflict when the Red Sea shores had been designated as restricted military areas.[56]

However, once the airlines started flying directly to the Red Sea, the tourism market increased exponentially with the real boom hitting in 1991. Again, incentives were given for investment which included ten years tax exemption.[57] Both local and international investors took advantage of these tax breaks to build and between 1989–2005 annual fixed investments in hotels increased from E£144 million in 1988–89 to E£274 million in 2004–5 or about $475 million with hotel capacity increasing from 40,000 to 170,000 rooms.

However, the quick construction of hotels and infrastructure led to 'over-speculation, over-construction, and low-quality construction' especially in Hurghada making this the cheaper place to go for a beach holiday.[58] Although the land along the coast had been earmarked for tourism little thought had gone into environmental and social impact. Also, over supply into a single investment has led to tourism self-destruction where tourism damages an area so much that it is no longer attractive to tourists. This phenomenon can be seen all over the world where, ironically, tourists do not want to travel somewhere as it is too 'touristy'.

However, beach holidays are not for everyone, and there are also opportunities for holidays working on an archaeological excavation. For example, the South Asasif Conservation Project in Luxor were able to fund an excavation with paying 'volunteers', as did the AERA-ARCE Field School Program in Cairo which started in 2008 and has now run seventeen projects which provide college credits for archaeology students whilst training them in the basics of archaeology.

For travellers who are less interested in antiquities and more interested in the modern culture, belly dance holidays became increasingly popular prior to the revolution. These present the experience of a different part of Egyptian life. Yasmina of Cairo,[59] for example, runs a belly dance B & B in Giza, and hosts dancers offering classes, a photoshoot at the pyramids as well as trips to see dancers perform at clubs and restaurants and of course the opportunity to buy professional costumes in the Khan el Khalili market.

For those who prefer folk dance and the rural aspects of Egyptian culture, an alternative is the Journey Through Egypt courses run by Sahra C. Kent, which includes an element of student group trips to Luxor and Aswan to learn more localised forms of dancing from local experts as well as providing 'cultural interviews, live music, dancing, historical walks, and cultural experiences'.[60]

However, even these holidays have had a detrimental effect on the lives of local Egyptians, as for the most part now, professional dancers in Egypt are Western. Ironically, they are preferred to Egyptian dancers as they are seen as 'exotic', meaning many of the Egyptian dancers are left dancing in less salubrious clubs for a lot less money than Western dancers charge.[61] In the 1990s there was a great deal of resentment against the influx of Russian dancers heading to Egypt,[62] and I remember seeing my first belly dance show and being told proudly by the staff 'she's Russian' which I thought was disappointing as I wanted to see authentic Egyptian dance.

## The Importance of Tourism in the Twentieth Century 21

The influx of foreign dancers to Egypt became a problem recognised by the authorities and in 2004 they were banned between January and September enabling Egyptian dancers to work. In 2022, there were twenty to thirty boats in Cairo which offered belly dancing shows with 'no Egyptian dancers ... working on the boats. Only foreigners ....'[63] To a certain extent this is because foreign dancers are able to subsidise their dream of dancing in Egypt with money from home, meaning many will dance for free or for 'pocket money' making them more attractive than Egyptian dancers who need a living wage.

Another form of 'adventure' tour is the day trip to Mansheya Nasir, also known as 'Garbage City', in Cairo. This is very much a symptom of the enclave tourism that has happened, separating the Egyptians from the tourists, but takes it to a darker place. This area of Cairo is inhabited primarily by Coptic Christians and is the place that the rest of Cairo dumps its rubbish. The inhabitants then scrape a living by recycling what they can. If the thought of people living in such poverty is not bad enough, it has become popular for rich Westerners to visit this poverty-stricken area with one blogger claiming, 'I managed to meet with Jaime ... and we both agreed that we *had* to see Garbage City.' He then described the area as well as useful tips on how to get there.

> Walking through the dirt streets gave us nothing but the crude reality of life in the slum. The smell of rotten food and decomposition floated throughout the city and engulfed us as soon as we stepped in.
>
> The dirt roads overflowed with stagnant water at some points, a visible sign that this slum lacks infrastructure and often has no running water, sewage, or electricity.
>
> In other parts where the dirt roads were dry, they simply were full of huge bags of garbage or simply garbage sprinkled all around. Animals like cows, goats, and dogs live inside buildings and on the streets, surviving on edible garbage.[64]

Is this a 'real' experience of the local culture, or is it voyeurism? The travel company Viator justify their day trip to Mansheya Nasir with a jolly description of the people living there:

> Its inhabitants are affectionately known as 'Garbage People', because they have been sustaining themselves for decades as Cairo's unofficial sanitation crew. The Garbage People, comprising mostly Coptic Christians, first immigrated to Cairo in the 1930s in search of big city dreams and land under their feet. Being very poor, and without a solid 401k plan, they got

by whatever way they could: living in makeshift housing and recycling the city's compostable waste products by feeding it to their pigs. Today, Cairo's Garbage People make ends meet by going door to door to pick up people's trash for a nominal fee, which is then brought back to their Garbage City neighbourhoods. There, they sort and recycle everything they could, using it for their own manufacturing.

There are numerous tour groups that take people for approximately £80 per person to this slum area of Cairo.[65] Considering that someone 'lucky' enough to work at the tombs in Luxor is earning E£100 a month (about £5), or a cleaner in a hotel E£85 per month (about £4), these companies are making a lot of money taking rich Westerners to gawk at the poor. It smacks of Victorian freak show sentimentality, and unless these tour companies are feeding that money back into the community to help with infrastructure, it seems immoral.

**Sex Tourism**

Another form of tourism which is not government led but which brings tourists back to Egypt multiple times a year is the Luxor/Cairo/Red Sea bride trend. This trend sees a young man marrying a much older Western woman and as Muslim men can have four wives, they often also have an Egyptian wife and family. Take for example 80-year-old Iris, and 35-year-old Mohammed. Iris's £200 a week pension is not a fortune in the UK where she lives, but to Mohammed who gave up his job before meeting her, it is a lot of money in the Egyptian economy.[66]

Typically the Western bride visits Egypt a couple of times a year, and throughout the rest of the year will provide money and gifts to set the 'husband' up in a business[67] or to help repair his felucca, kalash or house. Sadly, there are so many personal stories where the older woman (in her 50s, 60s or older) is flattered that a young man in his 20s or 30s wants to marry her and often emotions are generally one-sided. Iris, for example, is clearly besotted with her new husband and said: 'He's amazing in every way ... almost. His English is not amazing, and he can get words mixed up, but who needs language when you are in love?'[68]

These brides can then feel hurt when their husbands take an Egyptian wife and have children, effectively treating them like mistresses when they visit. I personally spoke to one German woman who was not able to stay with her husband in his house with his family when she visited Luxor, and as Egyptians are not allowed in the hotels, they would sleep on his felucca on the Nile for the duration of her visit. Sadly, this is all too common, and one bride said: 'I know women who have lost everything and then their Egyptian husbands left to marry the next foreigner.'[69]

There are also other stories of brides who have made the move to Luxor in their retirement years only to effectively live alone while their husbands live with their other families, and these brides pay for their husbands' businesses, children's' education and overall lifestyle until the brides' pensions are depleted.[70] Of course, there are many genuine marriages between Egyptians and non-Egyptians (which are on a more equal footing). Iris and Mohammed were still together in April 2022, and he had obtained a three-year visa to live in the UK.[71] However, there is no denying there is a manipulative money-making trend as well. According to Dr Izat Ashmawi, the head of the research department on human trafficking in the Ministry of Family and Population in Cairo, such marriages were on the increase and between 2000–2001, they had risen by five per cent: in 2002 by 29 per cent, while 2003 saw a 23 per cent increase, 2004 a 33 per cent increase and in 2005 a 30 per cent increase. In 2010 there were 17,000 cases of young Egyptian men marrying foreign women.[72] Dr. Hamdi Abdul al-Adhim, an economic expert said: 'Men from Upper Egypt [*southern*] dream of the opportunities they can obtain from such marriages. They dream to marry such women to come back to build homes, and own cars.'[73]

This trend of forming holiday romances is not new, and for a while in the 1990s, Dahab, in the Sinai Peninsula, was considered a great place for Egyptian men to go to have sex with Western women who travelled looking for a 'holiday romance'. As one guidebook describes it:

> If lazing on the beach, stoned, is your idea of heaven [Dahab] is the place to be. The music and ambience reek of the 1960s, when Israeli troops started coming here for R & R, introducing the Bedouin to another way of life. Nowadays, the real Bedouin village of tin shacks and scrawny goats hides behind scores of restaurants and campgrounds, while local children wander beneath the palm trees selling culottes and camel rides. Visitors will stay longer than they expected (sometimes until their money or brain cells are gone) or find the whole scene so repellent that they leave immediately.[74]

Like many aspects of Egyptian tourism, the Bedouin are needed in Dahab for their culture and hospitality as they are attractive to tourists, but the Bedouin are kept separate from most of the tourist industry and tourist spaces. Unlike other resorts like Sharm el Sheikh, in Dahab many of the hotels were smaller and run by local families, which attracted a different type of tourist.

Dahab had a reputation for attracting young hippies who stayed a long time but did not spend much money – exactly the tourists Egyptian authorities did not want. Such tourists buy independently from local shops and restaurants and use street taxis rather than those promoted through the large hotels and although this money stays in the local economy rather than getting lost in global corporate bank accounts, it is seen as undesirable.[75]

Therefore, several restrictions were put in place in Dahab in an attempt to raise the level of tourism services by introducing licenses for transporting and guiding tourists, inspections on properties resulting in fines, and hygiene inspections on restaurants which often meant small businesses were priced out. Add to this the 'sex' police, who cracked down on any relationship between Egyptian men and foreign women outside of tourism services.

It was illegal for an Egyptian man to stay in a hotel or camp with a woman who was not his wife, and the police were known to break down doors to arrest the men.[76] This led to people drawing up fake marriage certificates, or even getting married. The same crackdown on sexual behaviour was not, however, being carried out on Egyptian prostitutes and foreign men in Cairo and this is thought to be because, 'the Egyptians involved in sex tourism in Dahab are men, and the result is a disturbing emasculation which strikes a chord'.[77]

For contrast, in the 1990s sex tourism in Cairo was big business, with tourists from the Gulf States championing the industry. In 1995, 24.7 per cent of tourists to Egypt were from the Gulf and whilst they are not all sex tourists, they were all attractive to the Egyptian tourism ministry as not only did they understand and respect the Egyptian culture more than many Western tourists, they also spent a great deal of money.

It is thought that the Egyptian government turned a blind eye to the sex tourism, 'because of possible diplomatic embroilments' as well as 'the lucrative prospects of hard currency'.[78] At one end of the sex-industry are Gulf Arabs throwing money at Egyptian belly dancers in nightclubs on Pyramids Road and at the other end, there are apartments in Zamalek which are rented by the hour. Prostitutes will often go from flat to flat from five in the morning until the nightshift starts in late afternoon. As prostitution is illegal and against Islam, some of the women will enter into a 'marriage' with summer visitors known as *zawag al 'urfi* which is a marriage without witnesses which ends in divorce at the end of the summer.[79] Gulf Arabs also come to Egypt looking for brides and poor Egyptian farmers will often 'sell' their daughters for a substantial dowry which can often be tens of thousands of Egyptian pounds.[80] However, there are few crackdowns on this element of sex tourism as there is a lot of money involved, with backsheesh (tips) being paid to keep everyone sweet.

Tourism has developed a lot since the middle of the twentieth century, with the advent of cheap and convenient flights, and the Egyptian government's drive to improve tourism. Not all the changes have been positive – for the monuments or for the people of Egypt – but they have been good for the perceived tourist experience, as well as for the income it brings to the economy. However, as mentioned at the beginning of the chapter, the general expectations of tourists were set in the nineteenth century when the concept of mass tourism was just starting and that is where we will go in the next chapter.

Chapter 2

# The Beginnings of the Modern Tourism Industry

Tourism as we think of it today started in Egypt in the nineteenth century and was cemented in the psyche of the well-to-do traveller with the introduction of Thomas Cook cruises which he himself described as 'a great event in the arrangements of modern travel'.[1] I doubt he really understood the truth in his words, and how his organisation would take travellers around the world until liquidation caused the firm to fold in 2019. The name, however, was so strong that the bankrupt organisation was bought out by Chinese firm Fosun Tourism Group in 2020 for £11m a year, and who run it with a fraction of the staff.[2]

The success of Thomas Cook's package tours in nineteenth-century Egypt was built upon a very strong tradition of the wealthy travelling independently in small groups. The increased numbers of independent travellers at this time made it clear that Egypt was a viable venture for the travel company. Amelia Edwards (1873) described many of the visitors who were in Egypt at the time and their reasons for being there:

> Here are invalids in search of health; artists in search of subjects; sportsmen keen upon crocodiles; statesmen out for holiday; special correspondents alert for gossip: collectors on the scent of papyri and mummies; men of science with only scientific ends in view; and the usual surplus of idlers who travel for the mere love of travel, or the satisfaction of a purposeful curiosity.[3]

That is a lot of eclectic people travelling to the same place, for a mixture of recuperation, adventure and research.

Egypt had become an important holiday destination because of Napoleon (1769–1821) and his team of savants returning to Europe from Egypt with documentation, stelae and statues, including the Rosetta Stone which led to the decipherment of hieroglyphics. This research demonstrated that Egypt was full of history, antiquities, culture and interesting flora and fauna making it a popular draw for explorers, would-be archaeologists and antiquarians. And in the nineteenth century, antiquities were big business.

With limited restrictions, wealthy Westerners could easily travel to Egypt and indiscriminately collect artefacts, statues and mummies to be sold on the antiquities market in Europe and the US. Bernadino Drovetti's (1776–1852) collection of inscribed papyri, for example, was sold to the king of Sardinia for more than £13,000,[4] the equivalent of £785,421 in modern currency. It is not surprising then that there were many agents working for collectors whose job was literally to travel to Egypt and export whatever would bring the most money.

## Respect for the Monuments

All travellers, not just collectors wanted adventure in Egypt, and up until the mid-twentieth century tourists were encouraged to scramble over the monuments and collect things to take home. Travellers' respect for the monuments was vastly different to the modern day and they were not protected to the same extent. Wallis Budge demonstrates this in his *Cook's Handbook for Egypt and Sudan* (1911)[5] as he offers advice for the average tourist for a day at the ancient sites: 'Shoes with stout soles, low heels and fairly wide welts form the best footgear to wear when clambering over the ruins of ancient temples and sites.'[6]

If such advice was not horrifying enough, many of the dragomen (guides) also provided their guests with hammers and chisels to facilitate carving their names on the monuments as a reminder of their visit. Amelia Edwards, a stout supporter of preserving the monuments of Egypt, also carved her name at Abu Simbel, but then complained at the damage caused by later visitors: 'I was told that our names are partially effaced, and that the wall paintings which we had the happiness of admiring in all their beauty and freshness, are already injured. Such is the fate of every Egyptian monument, great or small.'[7]

Perhaps she does not recognise the irony of complaining that her graffiti and vandalism are no longer visible due to damage caused by later tourists. Amelia Edwards, however, was not the only one to carve her name as a memento, as this had been a sign of having visited a site since the Graeco-Roman period (see chapter 6). However, the nature of graffiti had changed over the millennia as in the Graeco-Roman period much of the graffiti was a means of venerating the gods and monuments, but by the nineteenth century, it was about showing presence. Even at the time not everyone saw it as a noble venture with Gustav Flaubert, the French novelist who travelled to Egypt in 1849–50 commenting:

In the temple we read travellers' names; they strike us as petty and futile. We never write ours; there are some that must have taken three days to carve, so deeply are they cut in the stone. There are some that you keep meeting everywhere – sublime persistence of stupidity. ... One is irritated by the number of imbeciles' names written everywhere: on top of the Great Pyramid there is a certain *Buffard, 79 Rue Saint-Martin, wallpaper manufacturer,* in black letters.[8]

Whilst today tourists may visit the monuments and marvel at the graffiti on the walls, especially if they are of well-known people like Belzoni or Champollion and date to 1799 or 1850, they do not feel the same about graffiti dated to 2019 or 2020, making comments about mindless vandalism. It is fascinating that modern graffiti is viewed this way, but age gives the same act gravitas. Will modern graffiti hold that same importance in a century?

## Respect for the Natives

The respect many tourists held for the ancient monuments in that they were encouraged to climb all over them and remove bits for souvenirs, was second only to the lack of respect they had for the native Egyptians.

The travellers of the nineteenth century mostly held colonial attitudes and often looked for ways of demonstrating their superiority and their Englishness or Frenchness. This was achieved by viewing the Egyptians as 'exotic others' and comparing the natives' 'uncivilised' ways against their own 'civilised ones'.

European and American travellers wanted to experience the life of the 'exotic other' from the comfort of the Westernised dahebeya starting the practice of enclave tourism which is now the norm in modern Egypt (chapter 1). A dahebeya was styled on the thirteenth-century ceremonial barges of the Mamluk sultans and were between 12 and 30 metres, shallow and flat bottomed with two large sails. They could also accommodate oars if the need arose. They were unfurnished other than the basics, so travellers brought linen, crockery, lanterns, medicine, entire libraries, pianos and many other accoutrements from home that they simply could not live without.

The attitude to the natives is hinted at in Agatha Christie's *Death on the Nile* (1937) when Tim Allerton is described as different as he does not hold the 'ordinary Britisher's dislike, and mistrust of foreigners'. Other similar attitudes are reflected when Mrs Allerton does not want to be surrounded by Egyptian children because 'their eyes [are] simply disgusting and so [are] their noses'. Although fiction, Christie has captured the time she was writing about, and whilst this was set in the early twentieth century, the attitudes had not changed for some decades.

Many Western tourists in the nineteenth century believed native Egyptians to be little more than children and one guidebook advises that an Egyptian, 'should be treated with a kind but firm hand as if he were a child'.[9] It was thought to be acceptable to think this way and, therefore, to 'discipline' Egyptians accordingly. This idea of 'father/child' or 'master/servant' relationship was further perpetuated by the fact that it was unusual for a traveller to Egypt to have any interaction with natives which were not within the sphere of a service provider/client, so the travellers were only surrounded by Egyptians within their pay. And in fairness, the upper classes did not necessarily treat their servants back in Europe any better.

Although Western travellers felt and acted as if they were superior, they relied on native Egyptians for everything. As soon as a party arrived in Egypt one of the first things to be sorted was the employment of a dragoman or guide to organise the trip, accommodation and services. American journalist William Cowper Prime commented in 1855–6:

> The dragoman may be defined as the gentleman who travels with you. ... He becomes a part of yourself, goes where you go, sleeps where you sleep, you talk through him, buy through him (and pay him and through him at the same time), and in point of fact, you become his servant. But, if you choose otherwise, you may make him what he should be, a very good servant, and nothing more. He who cannot manage his own servants should stay home and not travel.[10]

For many European travellers, the contrast between European behaviour and that of the Egyptians was considered shocking and was often used to show the differences between 'civilised' (Western) and 'un-civilised'(Egyptian) behaviour. In his guidebook, Budge warns the visitor that:

> [he] will not admire the natives ... until he knows them fairly well ... the Egyptian in all classes is lazy, and he will never do more than he is absolutely compelled to do. ... Those who render him the smallest service will demand baksheesh [tip], as likewise those who render him no service at all, but who stand about, stare at him and obstruct the way ... it is only demanded by those who wish to be overpaid for their services and by beggars.[11]

However, in Budge's favour, he does call the Western tourists out on their poor behaviour when it came to interaction with the natives: 'Tourists should especially abstain from throwing money from the decks of steamers onto the landing stages or onto the banks of the Nile for the purpose of witnessing the scramble for coins; such exhibitions are mischievous as well as degrading.'[12]

## Nineteenth-Century Itinerary

Despite many of the problematic attitudes of the time, the tourists of this period defined the itinerary which was to remain in place for the next 150 years and is referred to by many as the 'Golden Age of Travel'. Nineteenth-century visitors travelled from Cairo to Aswan by dahebeya, following a strict itinerary which to a certain extent is still maintained on the cruise boats today. The only real difference between the nineteenth-century itinerary and the modern one is the timescale. A modern cruise is about speed and fitting everything into a few days, whereas a trip down the Nile by dahebeya could take as long as five months. The other main difference is that, for the most part, modern cruises do not traverse the whole of the Nile from Cairo to Aswan, and instead primarily run from Luxor to Aswan and Luxor to Dendera/Abydos, even though it is now possible to get a cruise from Cairo to Luxor. Ideally, a Victorian trip started in Cairo, in October, and ended back in Cairo in January or February which is still the popular season in Egypt as the temperature is easier for Western tourists.

Many travellers kept prolific diaries of their trips and activities or wrote extensive letters home providing a wonderful insight into a typical itinerary of a trip on the Nile. In addition to such written records, many travellers also created numerous sketches and watercolours of the sites visited.

However, a more unusual record of a Nile cruise itinerary from Beni Hasan to Aswan can be found in the records of Alice Lieder who travelled to Egypt in the winter of 1851–52. What makes Alice's records unusual is that there is no existing diary, although it is more than likely that she kept one. Instead, her journey is recorded through her paper squeezes.

People travelling in the early nineteenth century were unable to take photographs as, although cameras existed, they were expensive and not particularly portable. Although the daguerreotype camera was introduced in 1839, it was not until 1888 that the amateur roll film cameras were introduced enabling every tourist to take instant photographs.

To counter this, photographic studios opened in Cairo and later Luxor in the 1850s, which meant tourists could go and choose souvenir photo albums, with images from a stock collection. These in themselves offer a wonderful snapshot of what the monuments looked like at a particular point in history, as well as offering an easy option of a souvenir of their trip.

However, there was an alternative to buying expensive photo albums. Artistic travellers created drawings or paintings of the monuments and others took paper squeezes. Paper squeezes were a kind of paper mâché and provided a perfect imprint of a carved inscription.

To make a squeeze, absorbent paper was soaked in water and once saturated was laid over an inscribed wall and hammered onto the inscription using a hog-hair brush ensuring the damp paper was pushed into all the nooks and crannies of the carving. Once the paper was dry (which did not take long in the Egyptian climate) it was removed from the wall producing a paper squeeze, or a 1-2-1 sized impression of the carving. These were easy to make and easy to transport as they were light weight, and once back in the West they could be used to make plaster casts of the inscriptions.

Squeezes as a method of recording, however, have a bad reputation, and damage to colour on the monuments is often cited as a reason they should not have been used – and, more strangely, why they should be ignored today. However, damage caused by squeezes is greatly exaggerated. For example, Julie Hankey in her biography of Arthur Wiegall boldly states: 'Some had taken wet squeezes – that is, they had pressed wetted paper onto the paintings so as to take coloured prints – draining the colour from the originals and leaving streaks and smears behind.'[13]

In my PhD research, I studied 339 examples of paper squeezes. Unfortunately, thirty-seven were mounted so observing the rear was impossible and of the remaining 302 squeezes only thirty-one had visible paint traces. Of these thirty-one (nine per cent) with colour traces, are small and fragmentary in nature so Hankey's claim of taking colour prints is not borne out by the evidence. Only four colours survive in the samples: blue, yellow, black and reddish brown. The latter is the most common and was generally used for male skin-tone and survived in 48 per cent of the samples indicating that some pigments were more porous than others and prone to transfer – again not consistent enough to produce a print. Of course, such a small sample cannot be considered representative of all squeezes in existence, which run into the hundreds of thousands, but indicates that stripping colour from the walls is not as widespread a problem as we are led to believe.[14]

The tomb of Sety I (KV17) is often used as a case study of the level of damage caused by squeezes. However, it is often glossed over that the tomb of Sety I is unusual as it was a particular draw for early cast and squeeze-makers due to the delicacy of the carvings.

The ARCE (American Research Centre in Egypt) Conservation Project on the tomb records how Giovanni Battista Belzoni (1778–1824) alone, in a period of only two years,[15] took 650 squeezes here; 510 of beeswax and 140 with a mixture of beeswax, resin and vegetable fibres.[16] The damage caused by later paper squeezes was the least of their worries when conserving the scenes as paper fragments could be easily removed (comparatively speaking) whereas the brown stains and drippings from beeswax and the corrosion of the paint from gypsum was a lot more problematic and irreversibly damaging.

The very fact that some scenes in the tomb of Sety I were squeezed hundreds of times with varying materials including gypsum, wax and plaster, all of which are more damaging than wet paper squeezes, makes this tomb an anomaly rather than the norm.

Whilst there can be no doubt that applying wet paper to an ancient painted surface is harmful, my own study of squeezes show that comparing the squeeze with the original monuments (other than Sety I's tomb) show little if any fading of the colour on the wall.

Alice Lieder's squeeze collection is particularly fascinating as at every site she visited on her trip in 1851–1852 she took a series of squeezes in varying sizes, marking the date and location giving us a pictorial record of her trip. This enabled us not only to see which temples she visited, but also which images interested her, and more importantly, those which were easily accessible, giving some insight into the state of the monuments at the time.

Her squeezes show that, as was traditional, she travelled by dahebeya from Cairo straight to Aswan without stopping, and then with the current turned around and travelled north back to Cairo, stopping at the monuments along the way.

Alice visited her first monument on 25 December, Christmas Day, 1851 when she went to Dhakka, in Lower Nubia. Then there was a whistle-stop tour, spending one day at most monuments, but five days in the Luxor region:

- 27 December 1851 – Beit el Wali
- 28 December 1851 – Debod
- 1 January 1852 – Kom Ombo
- 3 January 1852 – Edfu
- 5 January 1852 – Tomb of Paheri, El Kab
- 10 January 1852 – Abd el Qurna
- 12 January 1852 – Karnak
- 13 January 1852 – Valley of the Kings
- 14 January 1852 – Karnak
- 15 January 1852 – Deir el Bahri
- 16 January 1852 – Karnak and Luxor Temple

- 17 January 1852 – Dendera
- 18 January 1852 – Abydos
- 24 January 1852 – Beni Hasan

How she returned to Cairo or Alexandria from Beni Hasan in Middle Egypt is not recorded. She either travelled back to Cairo without stopping, or did not make any squeezes of the stops or the squeezes that she did make from this part of the trip are lost.

Another aspect of paper squeezes to consider is that they are only possible to make with carved inscriptions, and there are many sites like painted tombs and pyramids which do not have carved images and could explain the paucity of squeezes north of Beni Hasan.

However, the squeezes, coupled with the diaries of other travellers at the time show a standardised itinerary of a cruise as well as the potential time spent at each site.

## Travelling in the Nineteenth Century

Alice's itinerary, to a certain extent, looks like a modern traveller's itinerary to Egypt, but the experience of nineteenth-century tourism was very different. For one thing, timings were not as strict because people holidayed for a season rather than a fortnight, so staying longer at a site was possible if desired. There also were no opening and closing times or ticket sales. These sites were part of the landscape and open to visitors all the time. However, getting from the dahebeya or the later steamer to the sites was not as straightforward a case of jumping in a coach or taxi as it is in modern times. The road infrastructure was not designed for tourists and, to a certain extent, had not changed much since the pharaonic period (Introduction). In fact the landscape would have been different to the one we see in the twenty-first century.

Before the first dam was built at Aswan in 1899, the Nile flooded every year meaning some roads were inaccessible between June and September. For example, the Nile flooded right up to the pyramids of Giza and swirled around the feet of the Colossi of Memnon in Luxor which, to a certain extent, would dictate when tourists could access certain sites.

Out of flood season, travelling to the pyramids at Giza was not straightforward and was a two-day trip by donkey from the river. The first road to Giza was not built until 1869 and a bridge across the Nile was not constructed until 1871.

Everything required for the trip including food, water, candles, paper for squeezes, dynamite for archaeological exploration, tents, beds and everything else a nineteenth-century traveller could not live without for two days, had to be carried across the desert. Naturally, the rich Western travellers did not carry any of these things themselves and instead hired teams of Egyptian servants to facilitate their journeys.

Once reaching the pyramids at Giza, the first port of call was to climb to the top. *Baedeker's* guide gives advice on how to do this:

> The traveller selects two of the importunate Bedouins by whom he is assailed, and proceeds to the N.E. corner of the pyramid, where the ascent begins. These strong and active attendants assist the traveller to mount by pushing, pulling, and supporting him, and will scarcely allow him a moment's rest until the top is reached. As, however, the unwonted exertion is fatiguing, the traveller should insist on resting several times on the way up, if so disposed. The ascent may be made in 10–15 minutes but, in hot weather especially, the traveller is recommended to take nearly double that time. Persons inclined to giddiness may find the descent a little trying but the help of the Bedouins removes all danger.[17]

Eventually it was thought three Bedouins were better than two, one for each hand of the traveller and one to push from behind. A most undignified means of getting to the top. You should also consider that any women making the ascent did so in long skirts, corsets and heeled shoes.

Once reaching the top, travellers had the most spectacular views of the Giza plateau and, of course, they would set up the tables and chairs which had been brought up with them to have a cup of tea before making the journey down again. The Egyptian government banned climbing the Great Pyramid at Giza in the 1960s, after at least five climbers a year had fallen to their death or sustained serious injury, but as with many things in Egypt, a baksheesh in the right hand meant it was possible to continue climbing until the 1980s.

As the trip to the Giza pyramids was a bit of an excursion from the Nile, there was also the option to spend the night in the Great Pyramid, which many people were keen to do, from Napoleon to Dorothy Eady (Omm Sety) to Alistair Crowley. Each claimed to have had a spiritual experience with Paul Brunton (1898–1981) saying he felt the burial chamber was:

> peopled with unseen things … there was something abroad which I sensed as evil, dangerous. A nameless dread flickered into my heart … monstrous elemental creations, evil horrors of the underworld, forms of grotesque, insane, uncouth and fiendish aspects gathered round me and afflicted me with unimaginable repulsion.[18]

His imagination clearly got the better of him as he then claims some high priests materialised and threatened him before placing him into the sarcophagus.

In the early days of travel some of the tombs near the Great Pyramid were set up with wooden bedsteads so visitors could stay there, although they were expected to bring their own mattresses and bedding – on the back of the donkey.

In 1836, Mrs Sarah Rogers Haight and her husband recorded that their tomb was a 'tolerably comfortable parlour', and was furnished with carpets and mattresses.[19] A near-by tomb had been turned into a grilled meat restaurant by Monsieur François who stoked the fire with dried up bits of mummy which were also used by locomotive drivers as fuel[20] and were later sold to the West for the same purpose. This disregard for the remains of the ancient dead was widespread and when, in 1900, they were building the Cataract Hotel in Aswan (see below) they uncovered more than 200 mummies which simply turned to dust when hit with their pickaxes.[21]

As tourism increased and access to the pyramids at Giza became easier and more streamlined, several must-do activities were introduced which have remained in place to the modern day. For example, R.T. Kelly, writing in 1902, stated that travellers at this time had to put up with a 'regulation camel-ride' and the 'constant irritation of importunate guides and curio-sellers', which was exhausting for the visitor and meant they were 'too tired' to appreciate where they are.[22] It was only in 2020 that camel and horse rides were banned at the pyramids, due to animal welfare concerns (see chapter 1).[23] Kelly also believed that mass tourism led to what we may today refer to as check-list tourism where visitors are led to visit certain sites out of necessity rather than a genuine interest. 'I rather suspect that were the truth known, most of the Nile visitors are secretly bored, and only "do" these sites under moral compulsion.'[24]

A common amusement for tourists included a little bit of 'archaeology' (and I use the term very loosely) as many liked to return home with some pottery, a shabti (humanoid figure which acted as a servant in the afterlife) or even a mummy and coffin if they were lucky. Many tourists came prepared and brought a small pack of archaeological tools. Guidebooks even provided practical information on how to find the best artefacts and how to obtain permission to remove antiquities from the country.

> On leaving the country luggage is liable to be examined, and no traveller should attempt to export Egyptian antiquities without a special authorization to do so.
>
> Antiquities should be submitted to the authorities of the Egyptian Museum, who will assess their value for export duty, and have them

duly sealed with the official seal, and will give the owner a signed permit addressed to the Mudir of the Customs, instructing him to allow the objects to leave the country.[25]

It remained possible to remove artefacts from Egypt in this way until 1922 following the discovery of Tutankhamun's tomb which saw a crackdown on amateur archaeology.

## The Dangers of Travel

With freedom of movement across the sites, and lack of an official tourist industry, travel in the nineteenth century was hazardous in many ways, even before tourists were able to put themselves in danger by climbing the monuments.

For example, in 1876 a dahebeya was overturned by strong wind, killing everyone on board other than one crew member and one passenger. Dahebeyas capsized regularly, although not always so fatally, and both dahebeyas and steamboats were grounded on a regular basis, despite having flat bottoms.[26] Amelia Edwards (1873) records the benefits of a dahebeya when dealing with being stuck:

> Of the comparative merits of wooden boats, iron boats and steamers, I am not qualified to speak ... We however saw one iron dahabiya aground upon a sandbank, where, as we afterwards learned, it remained for three weeks. We also saw the wrecks of three steamers between Cairo and the first Cataract ... the old-fashioned dahabiya, flat-bottomed, drawing little water, light in hand, and easily poled off when stuck, was the one vessel best constructed for navigation on the Nile.[27]

Although dahebeyas were designed for the Nile and its currents, if there were crosswinds, they needed to be towed by one of the steamboats or row boats.

In the early nineteenth century there was also a genuine threat from Nile crocodiles which swam freely in the river, although the threat was generally greater for locals than travellers. However, most of the dahebeyas had a gun on board which was used to scare off any crocodiles wallowing in the shallows.

Nevertheless, taking advantage of the situation, crocodile hunting was marketed as great sport for travellers and in 1869 the Prince of Wales (later Edward VII) claimed to have shot a female crocodile with eighty eggs inside her. However, as the tourists increased, the crocodile population decreased as a result of hunting for sport but also by locals who then sold the desiccated beasts to tourists as souvenirs. By 1880, and with the increase of steamers on the Nile, crocodiles were not seen north of the first cataract and by the time the first Aswan Dam was built at the turn of the twentieth century, there were no more crocodiles in the Nile.

## Thomas Cook Steamers

The real instigator of mass tourism in Egypt was Thomas Cook and he was the pioneer of modern Egyptian tourism. The company's first trip to Egypt was in 1869, when Thomas hired two English-built steamers, the *Benha* and the *Beniswaif* at a cost of £40 each (a modern equivalent of £3,200). A steamer could travel the length of the Nile in approximately thirty days, which was far quicker than the traditional dahebeya, which could take up to five months. These steamers were small by modern standards and carried approximately forty-four passengers.

Unlike travelling by dahebeya, the entire journey could be booked from England or Europe which was a welcome convenience for many travellers. Traditionally, one of the first jobs of a traveller when arriving in Egypt was to choose a dahebeya from the hundreds on offer on the riverbanks, along with a rais, or captain, and his crew.

Amelia Edwards recorded that for ten days, for three to four hours a day she was dahebeya hunting which she found exhausting. As travellers stayed on the dahebeyas for months they treated commissioning one rather like house hunting and negotiated everything from the cost of hire to the wages of the crew. Edwards describes a visit to Boulaq, where there were hundreds of dahebeyas moored by the Nile ready for hire.

The boats were generally designed along the same plan, but varied between six to ten cabins, some with kitchens and others without, and a lounge or parlour. Some of the larger dahebeyas, like that commissioned by Ms. M.L.M. Carey, was 29.5 metres long and carried twenty-three people (five passengers, the dragoman, rais, steersman, fourteen crew, and a cook).[28] The cost also varied from day to day, with captains seizing every opportunity to earn as much money as possible from tourists. Negotiations also had to be conducted through a dragoman as none of the captains spoke English. John Gardner Wilkinson (1797–1875) advised travellers to make it clear that the captain was not to take any other passengers or transport any goods for the duration of the hire.[29]

John Murray's *Handbook for Travellers to Egypt* was the first of its kind being published in 1847, and advised that the dahebeyas 'should be sunk before a voyage to rid them of rats' and 'other noxious inhabitants', which does take the luxury out of such trips.[30] However, it was even less luxurious if a boat had not been sunk before habitation as Bartlett describes: 'The scoundrel of a rais had neglected to sink the boat as he had promised and from every chink and crevice in the old planks hundreds came forth scenting the blood of an Englishman.'[31] Thomas Cook removed this painful start to the trip which appealed to people who wanted to travel purely for leisure and recreation, rather than to study, excavate or for adventure.[32]

The all-inclusive packages started in the UK, with Thomas Cook meeting a party at London Bridge and escorting them through Europe to Egypt where they boarded a steamer. The journey from London to Alexandria took seven days and cost about £20 sterling.[33] Once on board the steamer there was a tight schedule starting at 8am for coffee, with breakfast at 10am, lunch at 1pm, dinner at 6:30pm and tea at 8pm. There was also a strict itinerary for sightseeing with up to four days in Alexandria,[34] fourteen days in Cairo, and then twenty-one to twenty-four days on the Nile to Aswan, including three or four days stop in Luxor.[35] Like modern cruise trips, with one or two temples a day it is easy to get information overload and the early steamer cruises were recommended for 'a tourist who can endure an ordinary amount of fatigue'.[36]

Miss Riggs, an unaccompanied traveller from Hampstead in London, was on the first cruise in 1869 and records the schedule in her diary showing the speed that they covered the sites:[37]

- Beni Hasan – 3 hours
- Assiout (Asyut) – 5 hours
- Girgeh (Girga) – 2 hours
- Kenah (Qena) – 8 hours
- Looksor (Luxor) – 3 days
- Edfou (Edfu) – 6 hours
- Koom Ambou (Kom Ombo) – 2 hours
- Assouan (Aswan) – 2 days

The first cruise in 1869 was an exciting one, and for much of the trip, the *Benha* and *Beniswaif* sailed behind Edward, Prince of Wales, and the royal party's fleet of six blue and gold steamers, each towing a barge of necessities and luxuries. These luxuries included 3,000 bottles of champagne, 4,000 of claret, a white donkey, horses and taxidermist Samuel Baker, naturally, to preserve any animals hunted whilst on the trip.[38]

Thomas Cook, tried to stay a couple of days behind the royal party as the prices when they were in town were a lot higher than normal and he wanted a better deal for his clients.

Thomas Cook marketed his trips not only to men as was traditional, but also to unaccompanied women and promoted himself as 'the travelling chaperone'[39] and indeed one of his first customers was Miss Riggs, who sailed on the *Benha* and shared a room with Miss Porter from Palace Clogher, Northern Ireland. Miss Riggs kept a diary of the trip and recorded her disappointment at missing the royal party. However, she was kept occupied and had packed her riding saddle as she intended to participate in hunting trips.[40] It was possible to purchase a quail shooting license through Thomas Cook & Sons as well any ammunition required for the trip.[41] Hunting was clearly big business and one traveller recorded killing 10,000 birds in one season.[42]

Despite these being luxury tours, and only available for the upper-middle and upper classes, these first boats were infested with what Cook called 'F sharps' (fleas).[43] There was also the additional problem, especially when there was a low Nile that the steamers would run aground, and on the first cruise both the *Benha* and *Beniswaif* got stuck and had to be dragged to deeper water by the crew:

> Now we are stuck on a sandbank ... The crew start to their feet, seize the boat-poles, and sticking them into the mud, push away with all their strength, bending themselves double as they walk along the side of the boat one after the other, and then withdrawing them, return quickly to repeat the operation. It is of no avail, and three of them proceed to take off their clothes, jump into the water, and put their shoulders to the boat: and while their fellows continue pushing as before from the deck, they lift the huge weight until we float again.[44]

When Thomas Cook returned to Cairo in 1870, he was appointed agent for Nile Passenger Traffic, hiring the Khedive's steamers which after refurbishment were called Cook's Steamers.[45] Cook controlled all passenger steamboat traffic on the Nile as far as the first and second cataracts[46] bringing the company good profits. When John Mason Cook took over the business in 1885—86, he built a new Nile fleet of their own which Faulkner described: 'The new steamers will be floating palaces and will be finer than anything that has floated on the grand old river since the days of Cleopatra. They will cause a sensation and I hope will prove a great attraction.'[47]

The new fleet was popular and saw the start of mass cruise-boat tourism in Egypt. Whilst popular with some travellers, not all were happy, however, to see an influx of Cook's travellers. In 1910, Pierre Loti lamented:

> Poor Luxor! Along the banks is a row of tourist boats, a sort of two or three storeyed (*sic*) barracks, which nowadays infest the Nile from Cairo to the Cataracts. Their whistlings and the vibration of their dynamos make an intolerable noise. How shall I find a quiet place for my dahebeya, where the functionaries of Messrs Cook will not come to disturb me?[48]

This increase in tourists via Thomas Cook & Sons, was resented by those who had been travelling to Egypt for years and considered the Thomas Cook tourists as 'less cultured' than they were and were referred to as 'cookies' or 'cookites'.[49]

Amelia Edwards commented in 1873 that a visitor immediately, 'distinguishes at first sight between a Cook's tourist and an independent traveller'. She clearly thought a Cook's tourist was of a lower class. She was not alone. Captain John Ardagh (1840–1907) stated:

> The enjoyment of such a trip depends in very high degree on the individuals forming the party, and Cook's Tourists in general do not possess a very high character among travellers. ... We had four seasons, which we distinguished as follows: first, flies; second, mosquitoes; third, flying bugs; fourth, Cook's Tourists.[50]

However, Cook's tourists were far from the lower classes and included Lord Salisbury, and Joseph Chamberlain.[51] Many still thought the increase in mass tourism to be 'the decline of the traveller and the rise of the tourist',[52] and many very much objected to travelling in mass groups which was the mark of a tourist over a traveller.[53] It is clear the semantics discussed in the Introduction were considered important even at this stage. One member of the royal party commented on the destruction caused by tourists:

> Ours is probably the last generation which will be permitted to see the glory of Egyptian sculpture, as they were first revealed to the explorers at the beginning of the century ... the smoke of the travellers' torches and the disfigurement by travellers' names, and the injury by travellers' spoilations, have rendered the 'fine gold dim' in many of the paintings and inscriptions.[54]

Whilst he may well have been right in his observations, he has also fallen into the trap of seeing the destruction caused by tourists – of which he was one – but separating himself from them as a group. No one who travels to Egypt sees themselves as part of the problem. They are always separate from the 'tourists', as they consider themselves to be a responsible traveller.

By the 1880s between 5,000 and 6,000 Cookies had passed through the Thomas Cook offices,[55] and in the winter of 1889–90 nearly 11,000 tourists hit Cairo, with 1,300 travelling down the Nile. In the winter of 1893 more than 1,000 visitors came to Cairo in a single week.[56] Mass tourism was in its infancy, but it was clear that the trajectory was going to continue.

William Howard Russell (1820–1907), an Irish journalist, commented on this new type of tourism, with no limit to his despair at the situation:

> That is a nuisance to the ordinary traveller to have his peace broken, to have a flood of people poured into a quiet town, to have hotels and steamers crammed, to see his pet mountain peak crested with bonnets and wideawakes, to behold his favourite valley filled up with a flood of 'mere English, whom no one knows.' I am not prepared to deny; but what are we to say to 'the greatest good of the greatest number?' Let us reflect and submit. The people at Alexandria were, as far as I could judge, very respectable – it was only in the concrete they became disagreeable. Mr Monpensier Brown and Miss Clara de Mowbray may be the capital companions as individuals, in the abstract but as 'Cook's Tourists' they become an aggregate of terrors.[57]

What was so wrong with Cook's tourists? Were they getting drunk, destroying monuments and generally being an embarrassment to their country? Were they smashing up bars? Not really. Considering these were still upper-middle class tourists, they were behaving pretty much the same as the condemning travellers. The problem seemed to be the numbers rather than anything else, although Amelia Edwards did comment on the damage, they caused to monuments. However, Lucy Duff Gordon complained that mass tourism was making a commodity out of Egypt, hitting the nail on the head.[58]

Journalist Sir Sidney Low (1857–1932) decided to enlighten his readers on 'the clients of Cook' and the 'relations of some European lady-visitors towards certain picturesque Arab ruffians who swagger about in the capacity of dragomans'. Apparently, there may also have been holiday romances between European women and Egyptian young men in the nineteenth century, which is definitely a practice which has continued (see chapter 1), although it may have been a little less intimate then than now. However, such behaviour in the nineteenth century was scandalous. At this time, most visitors to Egypt had an imperialistic view of the world, and anyone who was not European, with European sensibilities was considered 'inferior.' Through this lens, such a relationship was indeed scandalous. This imperialistic attitude peppers the diaries of travellers to Egypt (both Cook's tourists and independent travellers) and makes for rather sobering reading.

Initially, the Thomas Cook steamers were all the same standard but as a means of increasing tourism in 1886 they introduced the Express Service for those who could not afford the luxury of a two-to-three-week cruise. Then, in order to cater to those where money was no object, they introduced steam-powered dahebeyas.[59] These harped back to the travellers in the first half of the nineteenth century, bringing with them nostalgia, but as they had all been refurbished, they also brought a lot of luxury too.

Egypt, however, was not the overall goal of Thomas Cook, and he wanted to travel 'to Egypt via China' which became a possibility at the end of 1869 and the opening of the Suez Canal. Thomas Cook organised a celebratory trip for the opening of the canal on the 17 November. Unfortunately, it was not as successful as he hoped and only thirty passengers booked onto the trip and each paid 50 guineas.[60]

The opening ceremonies were a lavish affair, although there were no British dignitaries in attendance other than an untitled representative from the foreign office. There was a parade of sixty-seven ships making the first journey from the Mediterranean Sea to the Red Sea with Empress Eugenie of France and Emperor Francis Joseph of Austria on the *L'Aigle* at the front of the procession. They were followed by yachts with Khedive Ismail, the prince royal of Prussia, and Prince Henry of the Netherlands. Thomas Cook's paddle steamer, *America*, was thirty-sixth in line and had been refurbished as a luxury hotel. All the ships had coloured lanterns hanging from their masts ensuring Port Said was ablaze with colour and light. Following the ceremony, Thomas Cook and his party travelled by train to Cairo, to attend a meal hosted by the Khedive, with free champagne which Thomas Cook claimed 'had no attraction for my little teetotal party'.[61] Cook had been part of the Temperance movement for many years. Upon arriving in Cairo, he chose to stay at Shepheard's Hotel, the main hotel for Westerners.

## Luxury Hotels

When travelling from one town to the next, the river was the quickest means of transport, but when at the main tourist hubs – Cairo, Luxor, Aswan – many travellers left their dahebeyas or steamboats and descended upon the local hotels as a means of meeting other travellers, picking up messages, exchanging money and generally being seen. This led to the opening of luxury hotels which were a destination in themselves. There were four main hotels which were the centre of the travelling scene, three of which are still open to guests: Shepheard's, Mena House, the Winter Palace and the Cataract.

## Shepheard's Hotel, Cairo

Shepheard's Hotel was opened in 1851 and was originally called the British Hotel. It was located on a site adjoining Cairo's Ezbekiyah Square which had been owned by Abbas Pasha (Mohammed Ali's grandson) and was filled with lush greenery and rare trees. It was also the residence which Napoleon commandeered when he charged into Cairo in 1798. When he left Egypt it became the headquarters of General Kléber who was assassinated in the garden in 1800.[62] The land was purchased from Abbas Pasha by Samuel Shepheard for two whippets called Bess and Ben.[63]

The hotel received its first guests in July 1851 and, sadly, it was destroyed by an explosion in January 1952 as part of the anti-British riots ending in Egyptian independence. Although it was rebuilt on another site, it never regained the reputation of its heyday.[64]

When it initially opened in 1851 there were a number of disparaging opinions with John Gardner Wilkinson (1797–1875) saying the food 'leaves a lot to be desired', American writer Mark Twain (1835–1910) describing it as the second worst hotel on earth, Edward Lear (writing in 1867) called it, 'a pig-stye mixed with a bear-garden or a horribly noisy railway station' and *Murray's Handbook* claimed it was bad but still better than the steamers. High praise, indeed, considering the early steamers were full of fleas. The bedrooms were said to be, 'wretched in point of comfort', with beds 'so small that you are sure to fall out by morning no matter how quietly you repose' and with the 'aspect of a grim old barrack than a hostelry'.[65]

As with modern day Trip Advisor reviews, not everyone felt the same and a British journalist writing in 1853 commented there were large quantities of soap and water, 'boundless supplies of sweet, fair Turkish towels' and that room service was a dream 'when I order a narghile and a cool glass of Hodgson's pale ale; and having been supplied with these luxuries, we shout for the latest numbers of *The Times*'.[66]

Samuel Shepheard sold the hotel for £10,000 in 1860 to Philip Zech, a Bavarian hotelier from Alexandria. Following a fire, the hotel underwent refurbishment in 1869, which still did not silence the critics, so in 1890 it was torn down and rebuilt in the Italian style with marble columns, chandeliers, elaborate carpets and muted lighting. The building surrounded a courtyard garden, in addition to further gardens with palms, fountains, pelicans and flamingos. One of the restaurants was decorated like a Faberge egg with diamond-studded columns and painted ceilings. The ceilings in the entrance hall had been copied from Egyptian tombs and were brightly coloured. It was a fabulous sight. There were 340 bedrooms and 240 bathrooms which was unusual as many hotels had one or two bathrooms per floor. Shepheard's was the first hotel in the Middle East to be fully electric, with each room boasting an electric bell for room service. It firmly put the hotel on the path to being the place to be seen that it became in the late nineteenth century.

> Shepheard's was the epitome of glamour, a hotel from which explorers set off for Africa, where kings entertained mistresses, where movie stars rubbed shoulders with officers on leave from the desert war, their uniforms still dusty with the sands of El Alamein – and spies hovered in the hope of minds being softened by the congenial atmosphere.[67]

The front terrace became the biggest hive of activity and was reached by the central stairs to the main entrance which was flanked by two sphinxes that had been removed from the Temple of Serapis at Memphis. The terrace flanked the staircase and enabled the guests to watch the world go by on Kamal Pasha Street, and see who was entering the hotel. It also allowed guests to be seen by people walking by. John Ripley recorded watching the tourists return, from the terrace of Shepheard's:

> Costumes of every kind, complexions of every shade, and tongues of every nation, greet the eye and salute the ear: dragomans settling accounts with victimised tourists, vendors of photographs, shawls and jewellery, offering their wares; now one's attention is distracted by a juggler, who with carpet spread, is 'thimble rigging', and drawing long strings of medals out of his throat, before he finishes his conjuring tricks; another arrives with trained monkeys, dogs and goats, and solicits our attention to his exhibition; scarcely have we time to look at him when an Egyptian serpent charmer comes, and shows with what familiarity he can handle reptiles and turn them into neckties ... And thus the time speeds on to 6:30 pm, when the sound of the dinner gong disperses the artistes, and calls the visitors to the *table d'hôte*, and darkness and silence fall upon the city.[68]

By 1873, Shepheard's was a hive of tourist activity, and Thomas Cook & Sons had set up a tourist office in the grounds of the hotel, putting the hotel on the map with numerous high-profile guests. Amelia Edwards described the dining room in 1873:

Here assemble daily some two to three hundred persons of all ranks, nationalities and pursuits: half of whom are Anglo-Indians homeward or outward bound, European residents, or visitors established in Cairo for the winter. The other half, it may be taken for granted, are going up the Nile. ... The new-comers' first impulse is to inquire from what motives so many persons of dissimilar tastes and training can be led to embark upon an expedition which is to say the least of it, very tedious, very costly, and of an altogether exceptional interest.[69]

The hotel went through various upgrades in 1904, 1909 and 1927 to ensure it remained fashionable and modern. This included, following the discovery of the tomb of Tutankhamun in 1922, the conversion of the Louis XVI ballroom into a Pharaonic Hall.

Even in the later history of the hotel it was a special place to be seen. In the Second World War it was used by allied troupes who were entertained by dancer Josephine Baker as part of the Entertainments National Service Association (ENSA). Performers gathered at Shepheard's and it was where Noël Coward bumped into Josephine and said she was 'looking really the last word in chic'.[70]

However, by 1952, anti-British sentiment was high in Cairo, and on 26 January a mob gathered outside the hotel, piled up the wicker furniture from the terrace, poured petrol on it and set it on fire with all the guests still inside. They managed to escape before the mob entered the hotel and started destroying it.

When I got [to Shepheard's] I found a huge mob of people shouting and waving their arms as they watched the blazing remains of Egypt's most famous hotel collapse. Each time a timber crashed, or a wall fell they cheered. On the third floor a middle-aged woman stood at a window. She was holding a small bundle. When her room started filing with smoke she dropped the bundle out of the window and fell backwards into the room. I heard someone say the bundle was a baby.[71]

Only two bodies were recovered from the remains of the hotel.

Today, El Gomhoreya Street shows no sign it once housed the most famous hotel in Egypt.

## Mena House, Cairo

Outside of central Cairo, overlooking the pyramids at Giza, was the Mena House Hotel which was popular with health tourists due to the clean, dry desert air. Mena House was originally a hunting lodge, built by Ismail, the ruler of Egypt in the 1860s. In the 1880s, it was sold to an English couple Frederick and Mary Head, who extended it, but who only used it as a private residence to entertain guests. They named it Mena House.

They then sold their extended lodge to Hugh and Ethel Locke-King, another English couple, who hired architect Henri Favarger to build a luxury hotel next to the lodge. They intended to live in the lodge and save the hotel building for guests. The lodge was destroyed in 1950 as part of a further hotel expansion. It was sold at the beginning of the twentieth century to the Egyptian Hotels Company, who also ran the Savoy, Continental and Helwan hotels. Its final sale was in the 1970s to the Oberi group, an Indian hotel giant, which was run by Mohan Singh.[72]

The Mena House Hotel was built with recuperation of guests in mind, so all the main bedrooms face south and the sun, rather than the north and cool breeze (FIG 5). It opened in 1886 with eighty guest bedrooms and followed a traditional Arab style (rather than Pharaonic) with mashrabiya-style windows and architecture inspired by local mosques.

It was designed to be homely and included log fires and milk, fresh from the hotel's own cows. It also ran a shuttle service to Cairo which comprised a four-horse coach which ran between the Thomas Cook office and the hotel and took approximately 90 minutes.[73] The first swimming pool of its kind in Egypt was added in 1890 and was filled from a local spring. By 1900 a pyramid tramline had been installed which connected the pyramids to Cairo, meaning it was possible for people to make the journey to the pyramids and then to Mena House for lunch, returning to Cairo on the same day, which increased footfall to the hotel.[74]

As is to be expected there were mixed reviews of the hotel and the author of *Cairo of To-Day* said: 'If the pyramids had to be vulgarised, they could not have been vulgarised better (or less) by the English capitalist who is responsible for the Mena House.'[75]

However, it was this position overlooking the pyramids which was to make the hotel a major draw for travellers:

> It would be difficult to find a more delightful place to be idle in than the Mena. Those majestic masses that tower like mountains nearby seem to indue disinclination to movement. Then, too, sitting down on the desert,

with the pyramids for companions, it were impossible that any creature of average sensitiveness should not be conscious of the atmosphere of bygone ages which perennially abides about them.[76]

Although built to help visitors recuperate from illness, this did not mean it was a quiet, dull place. Amédée Baillot de Guerville (1869–1913) describes it thus:

> At the tea-hour its terraces are crowded with a gay and brilliant throng. The large and comfortable salons, the delicious Moorish dining room, the excellent food, the open-air swimming bath, the golf course, the tennis courts, the croquet lawns, all go to make a stay at Mena House one of the most pleasant incidents of a trip to Egypt. ... There are often sporting meetings, which are very popular. The camel races are particularly amusing. ... Last winter, a camel, furious at being passed, seized in his teeth the leg of the jockey of his more speedy rival and bit with fury.[77]

There were also several famous guests at Mena House. For example, Arthur Conan Doyle, the creator of Sherlock Holmes, stayed here with his wife who was suffering with tuberculosis for six months in 1895. He took part in many of the events organised by Mena House like golf, balls and desert rides, but also jumped on a Cook & Son cruise to Luxor and Aswan. The hotel has also been a base for filmmakers with Charlton Heston and Yul Brynner staying here during the making of *The Ten Commandments* (1956), Roger Moore when filming *The Spy Who Loved Me* (1977) and Mia Farrow when filming *Death on the Nile* (1978).

However, the Mena House Hotel also became an important location for diplomatic events, due to its placement technically in Cairo but outside of the city "both in Cairo and outside Cairo",[78] as well as being surrounded by desert making it secure. In November 1943, Winston Churchill, Franklin Roosevelt and Chiang Kai-Shek met to discuss means of ending the war. They each stayed in private villas in Giza and returned to the hotel for talks. In December 1977, further political talks were held there, and the peace treaty between Israel and Egypt was drafted.[79]

Mena House is still open to guests and has been rebranded as the Marriott Mena House. It is still marketed as a place for relaxation, but also as a business location with 'hi-tech business spaces'[80] and a short walk from the Giza pyramids.

## Winter Palace, Luxor

The Winter Palace in Luxor was built in 1905 and opened in 1907 to accommodate the increasing number of tourists to the city (FIG 6). It was built as competition to the Luxor Hotel, which had been opened by John Mason Cook as a health resort for tourists (see below). Thomas Cook employed a number of local people in his enterprises and his foundry and machine shop employed at least 300 people, and when steamboat staff, hotel staff, guides, donkey boys and luggage handlers are considered, thousands of local people were probably employed.[81] Thomas Cook & Sons also funded a hospital in Luxor where Egyptians travelled hundreds of miles to visit, meaning there was a strong dependency on the travel company being a success.[82]

For the inauguration of the Winter Palace on 19 January 1907, guests were treated to a picnic in the Valley of the Kings before returning to the hotel for dinner and an evening of bridge and mingling with other guests. Within a week of opening, word had spread, and they were fully booked.

The Winter Palace was described by French naval officer, Pierre Loti (1850–1923):

> The thing which dominates the whole town, and may be seen five or six miles away, ... a hasty modern production which has grown on the border of the Nile during the past year; a colossal hotel, obviously sham, made of plaster and mud, on a framework of iron. Twice or three times as high as the admirable pharaonic temple, its impudent facade rises there, painted a dirty yellow. The old Arab town, with its little white houses, its minarets and its palm trees, might as well not exist. It is the end of Luxor.[83]

Despite such opinions, this hotel did not spell the end of Luxor, and this hotel is still one of the grandest in the town more than a century later. However, mass tourism did change Luxor (and the rest of Egypt) as more locals in search of work moved to the areas where there were tourists, starting the dependency on tourists for an income that was to prove so damaging in the later twentieth century.[84]

The Winter Palace in the early twentieth century was the place to be seen in Luxor and it was an important hotel for the rich and influential. Writer Douglas Sladen (1856–1947) described guests as 'splendid specimens of young Englishmen', and 'delicious and deliciously dressed English girls'.[85] However, it was not just the resting place of the holidaying English. It was also a regular spot for the international teams of archaeologists who were working on the west bank of the Nile. They would head to the Winter Palace for drinks, dinner and conversation, but also as a means of receiving and sending messages. For example, it was here in 1922 that Howard Carter announced the discovery of the tomb of Tutankhamun to the world, via a sign on the notice board. He was a regular customer, even with accommodation on the west bank, and until his death in the 1960s, he could be seen having tea in the gardens. It was here in 1933 that Carter met Agatha Christie and her archaeologist husband Max Mallowan when they were staying in Luxor, and they passed the time playing bridge at the hotel. The couple continued their journey to Aswan staying also at the Cataract. It was on this trip that Agatha wrote a short story *Death on the Nile* which was published in 1934 in a collection called *Parker Pyne Investigates*. In 1937, she then wrote the more famous novel of the same name. The original short story has never been adapted even though it has all the characteristics of a classic Christie, with a detective, suspicion, limited suspects, a love triangle and, of course, a murder. Also on this trip she started writing the play *Akhnaton*, although this was not published until 1973 and has never been a stage production.[86]

The discovery of Tutankhamun's tomb gave tourism to Luxor a major boost as people flocked there with the hope they would witness something wonderful. By December 1922, the Winter Palace was fully booked, and they set up tents in the grounds where the wealthy guests slept on army cots. Guests thought this bit of discomfort was worth it as they were all given tours of the tomb's antechamber, but not the burial chamber. This did not happen until the discoveries were officially announced in February 1923. The guests were at the centre of all the press and hub bub the tomb had generated.

The hotel is still open to the public, and in the 1970s the New Winter Palace was added to the grounds but was then demolished in 2008 as part of the scheme to make the area surrounding Luxor Temple more aesthetically pleasing. It is currently run by the Sofitel Group, and, to a certain extent, maintains the splendour the hotel held in its heyday.

## Cataract Hotel, Aswan

The fourth hotel of note was the Cataract Hotel in Aswan which was built in 1899 by Thomas Cook to accommodate his own guests. It was designed by the architect Henri Favarger, who had been instrumental in designing Cairo's Mena House. The first advertisement in 1899 in *The Egyptian Gazette* stated it had: 'Every modern comfort. Large and small apartment rooms, library, billiard room etc. ... fireplaces in hall, salons and the main rooms. Electrical lights running all night. Perfect sanitary arrangements approved by the authorities. Can accommodate 60 visitors'.

It opened in January 1900 and was primarily a health resort as the hot, dry Aswan air was thought to be beneficial for many conditions. There were 120 rooms, most of which faced south to make the most of the sun, with balconies overlooking the river Nile. However, demand was so high that in 1901, tents were needed in the grounds to cope. They kept extending the hotel until, by 1902, it had another floor added with an additional sixty rooms, which brought the total up to 220.

W.E. Kingsford in his book, *Assouan as a Health Resort* (1899), described the benefits of the resort:

> In the construction of this hotel, great attention has been given to the requirements of the invalids – most of the rooms have verandas, and a warm, sunny aspect; many are fitted with fireplaces, and the position and form of the building has been chosen to provide shelter from the prevailing winds.
>
> The sanitary arrangements have been carefully studied, Moule's earth closet system[87] being adopted, and the water filtered through Reeve's gravity, and Berkefeld filters.[88]
>
> Every modern convenience is provided for in the form of electric light, hot and cold baths, &c., and a ... number of private sitting rooms to meet the requirements of invalids.
>
> There is an English physician and nurse in Assouan, and an English house-keeper is in charge of the domestic arrangements of the Hotel.[89]

The Cataract Hotel quickly grew in popularity with *The Telegraph* claiming it as 'unmatched even in Europe'. However, although it was popular with people who were there for their health, it did not stop the hotel from putting on a lot of events such as croquet, tennis, golf, gymkhanas, donkey and camel races, paper chases and even a polo match where one team had to be content with billiard cues as there were not enough polo sticks. Amédée Baillot de Guerville records in 1906: 'Hundreds of donkeys arrived at the gallop, donkeys of every colour and size, mounted by the most varied types of riders that one could meet with under the same sky. Everyone had come for a paper chase, the start and finish of which were to take place at the Cataract.'[90]

Other than a 'brutalization of the grounds in the 1970s',[91] the Cataract looks very much the same as it did in the 1930s when the last major renovations were done. The Cataract Hotel is still trading as a Sofitel Hotel, and is now called the Old Cataract Hotel to differentiate it from the new wing (The New Cataract) which was built in 1961. The hotel underwent some major renovations in 2011 where the new and old wings joined.

## Health Resorts

Egypt was a popular destination in the nineteenth century for the infirm and those recovering from illness. The hot, dry climate was thought to be good for wealthy Europeans, and from the end of the eighteenth century it was a very popular health resort. As the summer months in Egypt were too hot for most tourists they mainly travelled there in winter, which worked well with Thomas Cook's European Tours which only ran through the summer months. This meant he was able to create a year-round business.

There was a total of six health resorts in Egypt: Alexandria, Cairo, Mena House at Giza, Helwan, Luxor and Aswan. Each offered a different health benefit. Alexandria, for example, offered healthy sea breezes, with a drier and warmer climate than Cairo.

Only a short ride of 16 miles south of Cairo was the town of Helwan, where the sulphur and salt springs were believed by many to have healing properties which were 'most beneficial for [the] sick and suffering'.[92] The sulphur springs were compared to those of Aix in Savoy.[93]

> The air of Helwan is clean and free from sand and dust, and the restfulness of the place is very grateful; from the middle of November to the middle of April the climate is most beneficial for the sick and suffering. The baths which have been built during the last few years leave little to be desired, and it is not to be wondered at that it has recently become the fashion for the inhabitants of Cairo to resort there. The springs have been found specially beneficial in the various forms of skin disease to which residents in so hot a climate are subject.[94]

Luxor in the south of Egypt, however, was seen as the best place to visit for recuperation as it was less windy than the north, with constant sunshine and warmth, which was 'extremely grateful to delicate folk'.[95] Prolific writer on Egypt, Lucie Duff Gordon, travelled to the country in the 1860s to recuperate from tuberculosis. She died there in 1869. She was not much interested in the monuments and was literally there to take the air and to bask in the dry heat.[96]

The Luxor health resort was initially the Luxor Hotel, which was built by John Mason Cook in 1878, and was a sanatorium for pulmonary diseases. It was thought invalids benefitted from the dry air, and there was a doctor available on site for emergencies. In its second season, a new wing was added, doubling the capacity to forty-five bedrooms.[97] It cost 15 shillings a day to stay there, which John Murray thought was expensive, stating the accommodation and the food was substandard and would not benefit the unwell.

The hotter and drier climate in Aswan was considered even more beneficial for health, and many hotels were built to accommodate health visitors including the Cataract Hotel, which opened in 1900. Although it was seen as a health resort for recuperating 'invalids', the hotel still boasted all amenities and the first advertisement in *The Egyptian Gazette* (1899) praised the fireplaces, electric lights and sanitary arrangements.

Many people visited Egypt for their health including George Herbert, 5th Earl of Carnarvon, who chose to bank roll Howard Carter, resulting in the discovery of the tomb of Tutankhamun. Lord Carnarvon initially visited Egypt following the advice of his doctors after a car accident which injured his chest. His was only the third car to be licensed in England and he was seen locally as a bit of a 'boy racer' as he imported and drove the earliest cars. He also encouraged his friend Geoffrey de Havilland to make an aeroplane flight from the Highclere estate, no doubt to the horror of the locals. He visited Egypt for the first time in 1898 and when he returned in 1905, he stayed for three months and applied for an archaeological concession. He returned every year to excavate (other than during the First World War) until 1922 and the discovery of Tutankhamun.[98]

Even if they did not initially travel to Egypt as health tourists, as many visitors may never have left the safety of their European cities, the different food, water and unrelenting heat could have affected travellers already delicate constitutions. As common then as today was 'Pharoah's Revenge'. It is thought that illness was made worse by following the tourist trail rather than going on more unknown journeys. It was believed that the tourist trail often led to filthy fly-ridden campsites, which brought their own health risks.[99] Travellers were advised to carry: 'Warburg's tincture and quinine for fever; bicarbonate of soda, ginger, bismuth, for stomachic problems; cascara sagrada, and some aperients salt, chlorodyne, and a small quantity of tincture of camphor or of opium, for diarrhoea, and ipecacuanha wine for dysentery'.[100]

However, despite these precautions most travellers staying in a hotel or on a cruise boat, were 'more or less ill', also suffering the effects of the heat with sunstroke. In fact, sunstroke became such a concern that it was covered in the guidebooks, and Murray recommended wearing wool next to the skin, since linen 'checked' perspiration, resulting in 'fever or diarrhoea'.

## Souvenirs

We tend to think of Victorian travellers as vastly different from ourselves, when in fact they were not really. One way this is made clear is in tourist souvenirs. Granted, Victorian souvenirs were very different to their modern counterparts. In modern Egypt, some statues are made by local artisans with many more made of resin and shipped in bulk from China, whereas in the nineteenth century many of the tourist souvenirs were genuine artefacts in the form of statues, mummies or papyrus.

French archaeologist, Auguste Mariette who founded the Services des Antiquités, commented in the 1870s that 'we have no advice to give those travellers who wish to buy antiquities and take them home as souvenirs of their visit to Egypt. They will find more than one excellent factory at Luxor. But to travellers who really wish to turn their journey to some account, we would recommend the search after papyri.'[101]

This in itself seems dubious advice from an eminent Egyptologist, but everyone wanted something real and old. For example, travelling in 1869, the Prince of Wales (later Edward VII) collected a sarcophagus, thirty mummies, a live black ram and a ten-year-old Nubian boy to take home to England.[102] Lucie Duff-Gordon who travelled to Egypt for her health took original artefacts home to her friends and family as gifts.[103]

Whilst taking tea on the terrace at Shepheard's the guests were bombarded with an array of goods for sale:

> they may examine and buy Syrian picture-frames, ostrich feathers, bead necklaces, fly-switches, hippopotamus-hide sticks and whips, lace, braces, beans, pastry, suspenders, tarbrushes, air-balloons, birds in cages, roses, narcissi, carnations, hyacinths, coat-stretchers, Indian boxes, and when they are on the market, leopards and boa-constrictors ... .[104]

R.T. Kelly warned visitors of the industry within Egypt of the fake artefact trade, where it was 'surprising to what trouble and expense the native will occasionally put himself in their manufacture'.[105] Even Agatha Christie, in *Death of the Nile*, describes the experience at Aswan with the curios sellers:

> They came out of the shade of the gardens on to a dusty stretch of road bordered by the river. Five watchful bead-sellers, two vendors of postcards, three sellers of plaster scarabs, a couple of donkey boys and some detached by hopeful infantile riff-raff closed in upon them. ...
> 'It's best to pretend to be deaf and blind', she remarked.
> The infantile riff-raff ran alongside murmuring plaintively: 'Bakshish? Bakshish? Hip Hurrah – very good, very nice ... '.[106]

Sellers were as relentless in the 1930s as they are in the twenty-first century and, to a certain extent, sold the same things: scarabs, jewellery, donkey rides, boat rides and other 'real' items made in a workshop that week.

Like modern tourists, the nineteenth-century visitors also wanted reminders of some of the things they had seen. If they were not accomplished artists, they had a few options. They could purchase watercolours locally; they could purchase photograph books from the photographic studio, or they could purchase (or make) paper squeezes.

The tourist industry in paper squeezes is in the early stages of research, but many surviving squeezes are made of the same material, dimensions and scenes in collections all over the world (including Bristol Museum, Leeds, Petrie Museum, Griffith Institute and the Smithsonian). Each of them gifted by different donors and originally owned by different individuals who had travelled to Egypt.[107]

It appears that an enterprising Egyptian was producing squeezes of popular images from tombs such as that of Khaemhat (TT57) from the Valley of the Nobles at Luxor and selling them to tourists. No precise details of this industry have yet been studied, so how much they charged tourists, and whether they were bought on site, made to order or touted around the bazaar is unknown.

It was this drive to own a bit of the past, whether in a photograph, drawing or artefact, that had initially brought Westerners to Egypt in the first place. The seventeenth and eighteenth centuries were all about treasure hunting, discoveries and adventures, which lay strong foundations for this more sedate nineteenth-century tourism and souvenir collecting.

Chapter 3

# Travel in the Seventeenth and Eighteenth Centuries

The eighteenth century saw a new approach to travel in Egypt as antiquarians and historians realised what rich pickings the historical remains of ancient Egypt were. So popular was Egypt, in fact, that Napoleon decided that part of his military campaign would include studying the culture, flora, fauna and historical monuments to preserve them for prosperity.

The findings were to be published as the *Description de L'Egypte* which was a great catalyst for increased travel and exploration in the early nineteenth century, the era that can be seen as the end of independent 'exploration' in Egypt. This series of large format folios provided the West with the first records of many of the sites in Egypt and, even 200 years later, is still hugely valuable as many of the sites no longer look the same or have disappeared completely since first recorded.

## *Description de L'Egypte*

Napoleon arrived in Alexandria on 1 July 1798 and took up lodgings in Alexandria, near Pompey's Pillar.[1] Although he was essentially there to colonise Egypt,[2] he arrived with an entourage of 167 scholars as well as 54,000 soldiers. Like most Europeans he had an understanding of Egyptian history, and he was not that far out when he said at the Battle of the Pyramids (21 July 1798): 'Soldiers! From the top of these pyramids, forty centuries gaze upon you.'[3]

No one at the time knew exactly how old the pyramids were, but they knew they were old. It is recorded that Napoleon spent some time alone in the burial chamber of the Great Pyramid whilst he was there, and apparently had an experience which left him rather shaken. He never spoke about it until on his deathbed when he claimed; 'No, what's the use? You would never believe me anyway.'

Napoleon's team of 167 scholars included three pharmacists, seven chemists, four mathematicians, three astronomers, one student, three writers, one economist, two artists, five mining engineers and three students, fourteen surveyors and seven students, fourteen civil engineers and two students, three powder makers and two students, three shipbuilders, three mechanical engineers, one former knight of Malta, thirteen polytechnic students, five other students, nine mechanics, two musicians, twenty-seven printers and three printer's wives armed with blocks of Latin, Greek and Arabic characters.[4] Their purpose was to record the geography, topography, zoology, flora, fauna, culture, society and history of Egypt. Conversely, they were also there to provide Egypt with a modern infrastructure, with many trained to build bridges, roads, and canals.[5] They travelled with a library of books and were well versed in current literature, maps and papers concerning Egypt.[6] One of the most notable scholars in the group was Baron Dominique Vivant Denon, an artist, diplomat, writer and close friend to Madame de Pompadour, the chief mistress of Louis XV from 1745–51. Also a close friend to Napoleon, Denon was asked to lead the intellectual expedition. The expedition headquarters were in Cairo at the Mamluk palace of Hassan-Kashif, which included a library from France and the only printing press in Cairo, forming the Institut d'Égypte.

The group of scholars travelled the whole country from Alexandria to Philae recording every place, monument, artefact and item of interest creating: 'one of the great intellectual and artistic achievements of the nineteenth century … Opened the eyes of Europe to the splendours of the monuments and customs of Egypt and gave inception to the science of Egyptology.'[7]

Many of the French who travelled to Egypt were familiar with Thebes (Luxor) from Homer, who described it as 'hundred-gated' and were awed when they arrived there:

> This illustrious city … enveloped in the veil of mystery and the obscurity of ages, whereby even its own colossal monuments are magnified to the imaginate, still impressed the mind with such gigantic phantoms that the whole army, suddenly and with one accord, stood in amazement at the sight of its shattered ruins, and clapped their hands in delight, as if the end and object of their various toils, and the complete conquest of Egypt, were accomplished and secured by taking possession of the splendid remains of this ancient metropolis.[8]

They worked hard and returned to France in 1801, two years after Napoleon had abandoned his campaign in Egypt. They had recorded thousands of monuments and artefacts with passion and attention to detail, and it is incredible to think that all the internal records (e.g. tombs) had been made by candlelight. The scholars were clearly passionate about the task they had been given. Vivant Denon records his time at Medinet Habu, on the west bank at

Luxor, and his passion for the artwork he witnessed is apparent. When the party were called to leave the site, he was reluctant to go:

> How could I, thus hastily, leave these precious curiosities? I begged with earnestness for a quarter of an hour and watch in hand, I was allowed twenty minutes: one person lit the way, while another held a torch to each particular object to which I directed my attention.[9]

Between 1809 and 1829, the *Description de L'Egypte* was published and eventually ran to nine volumes of text and eleven of plates and then expanding in a second edition to twenty-five volumes, with five alone devoted to antiquities.[10] The series comprised of 894 plates, with 3,000 drawings, and a volume of maps which constituted the most complete record of Egypt at the time. Whilst the *Description* had a massive impact on the imagination of Europeans, it was not the only thing that drew people to Egypt.

The French also took back with them a number of artefacts which included the zodiac ceiling from the temple of Dendera and the Rosetta Stone, which was then confiscated from the French by the British and saw the start of the race to decipher the hieroglyphic language.[11] In order to do this, there was a major antiquities drive with visitors flocking to Egypt to purchase papyri and inscriptions covered in hieroglyphics.

The Rosetta Sone was discovered as soldiers destroyed an old, dilapidated wall in the Delta town of Rosetta, just north of Alexandria. Fortunately, the Greek text was identified, along with two other scripts which indicated could be a stone of some importance. Once the inscription had been deciphered, it was discovered to have been produced on the accession to the throne of Ptolemy V Epiphanes in March 196 BCE. The text was written in three scripts so it could be understood by both native Egyptians and the Greek community (for a discussion on language see the Introduction). During the Renaissance there were attempts to decipher hieroglyphics, although they were doomed to failure. The general belief was that hieroglyphs by their very nature were not meant to provide but to conceal information, meaning they were only readable by the long dead. This idea tied in with that of the: 'sphinxes carved in the Egyptian temples represented the safeguarding of mystic dogmas from the ungodly masses by means of complex enigmas'.[12] The discovery of the Rosetta Stone was to open a whole new scholarly path, as well as open up the history of Egypt to the masses.

## The Legacy of *Le Description de L'Egypte*

The combination of the publication of *Description de L'Egypte* in the first two decades of the nineteenth century and the decipherment of hieroglyphs in 1822 put Egypt on the map and it was appended to the so-called Grand Tour of Europe which was popular in the seventeenth to the nineteenth centuries, with Italy as the focal point. This meant there was a constant stream of Europeans arriving in Egypt to look at the monuments, publish guidebooks and travel journals and to take a little bit of the past home with them. At this stage in Egypt's history, papyrus or, indeed, any artefact with hieroglyphs on, was greatly desired, even though most travellers could not read them. Many dealers, who were considering profit over history often cut up papyrus rolls into smaller sheets meaning they could sell one papyrus to more people. Of course this has made a fun jigsaw for Egyptologists and papyrologists over the centuries as they have tried to put these fragments back together. Another means of gaining inscriptions was to simply hack them off the walls of tombs and this is something that you will see in the tombs of the Valley of the Kings and Valley of the Nobles – inscriptions literally removed from the walls and shipped to the West. For example, in the Louvre and in Florence, the door jambs from the tomb of Sety I are proudly on display. At the time there were no controls, and with the increased construction of plants to process sugar cane, many monuments were removed by the Egyptians as foundations for the factories and even to be ground down into quick lime.

The trend of collecting artefacts was not a new one at this time as the interest in Egypt and its history and antiquity had been a longstanding one. As early as the ninth century excavators were hoping to find treasure in Egypt. For example, Caliph el-Mamun (786–833 CE), records coating the north face of the Great Pyramid with hot vinegar, hoping to crack the stone enough to allow him to enter. Clearly this was not going to work, so he took a battering ram instead and forced his way in. This entrance is still the one used by tourists to enter the pyramid today (FIG 7).

There was very little preventing treasure hunters throughout the centuries from removing items. Even Mohammed Ali, the ruler of Egypt from 1805 to 1848 who was focused on bringing Egypt into the modern world was happy to use the antiquities as quarries for new building works, as well as allowing travellers to take them away – either as gifts or purchases.[13] The laws on removing items from Egypt were not changed until the discovery of the tomb of Tutankhamun in 1922. Therefore, many items – with inscriptions or not – were removed in bulk in the preceding centuries, filling museums and private collections around the world. Other items were removed and subsequently destroyed.

Mummies were particularly popular, both human and animal, and Vitaliano Donati visiting in 1759–60 records purchasing a number of animal mummies at Medinet Habu in Luxor which included the mummy of a bird, a cat and a dog, as well as a bust of Isis at Koptos and some oil lamps with Christian and Coptic inscriptions. But Donati's main souvenirs were rocks and minerals and he returned to Italy with 224 samples including porphyries and different types of granite.[14]

Trade was not a major factor between Europe and Egypt but in 1586, John Sanderson, a British merchant, had an arrangement with the Turkish government to transport 254 kilos of mummy pieces back to England, where they were to be distributed to the apothecaries to be ground down and used as medicine to cure a whole array of illnesses including paralysis or abscesses.[15] This became a cure throughout Europe and, thus, more and more were required, which led to a supply and demand issue.

Entrepreneurial Egyptians made mummies from criminals, the unclaimed dead, and those with severe disease, burying or sun-drying them to age them before selling them to unsuspecting European travellers to meet the demand.[16] This was an old practice and Ibn Iyas (d. 1542) records a case under Sultan Al-Ashraf Barsbay (1422–1437 CE) where this had been carried out. This practice was banned in the eighteenth century, and the Ottoman rulers taxed the transportation of mummies, meaning they cost more to bring back to Europe.

There were some physicians and those trained in medicine who condemned the use of ground mummy, but they were not heeded by the masses. The royal surgeon Ambroise Paré (1510–1690) for example stated: 'But the case stands this, that this wicked kind of Drug, doth nothing help the diseased ... as I have tried an *(sic)* hundred times.'[17] He tried to prevent others from prescribing it, but the masses went with the hype. Although not everyone agreed that ground mummy could cure everything, many physicians travelled to Egypt to learn about medicine and botany.

The Venetian physician Prospero Alpini (1553–1616) went to Egypt in 1581 and stayed for three years studying botanicals. However, he was also fascinated by the pyramids at Giza and recorded the various passages that he could see when he visited. He recorded that the Ottoman viceroy of Egypt, in his pursuance of the gold within the pyramids, intended to fill them with gunpowder so as to expose the inner structures. Fortunately, the Venetian consul persuaded him not to do such a ridiculous thing.[18]

Treasure hunting in the medieval period (chapter 4) was all about gold and precious stones, whereas in the sixteenth and seventeenth centuries, treasure meant antiquities. The seventeenth century also saw the introduction of purchasing antiquities through commissioned agents. One such agent was Johann Michael Wansleben (1639–1679) who was commissioned by Jean-Baptiste Colbert, a minister of Louis XIV. He was told by the librarian Pierre de Carcavy:

> The main object of the voyages which the King commands Sieur Vansleb to make in the Levant is to seek there the greatest possible number of good manuscripts and ancient coins for His Majesty's library. Should he find also among those ancient monuments any statues or bas-reliefs by good masters, he will try to obtain them. ... He will make a collection of ancient inscriptions that he finds and try to copy them exactly as in the same language in which they are written, having them read and explained by some interpreter, should he not be acquainted with the characters therefore.[19]

He was told these items could be found in the cemeteries and tombs and indeed he sent back to France more than three hundred manuscripts, three mummies, various idols and a number of natural history specimens, all within a year of being in Egypt. He was pretty hands-on in his explorations and records being lowered into pits at Saqqara and removing mummified ibises with his own hands. He was also one of the first to record the city of Arsinoë and Antinoopolis. However, he did not sail further south than Girgeh in the Sohag governorate, just north of Qena.

## Giovanni Belzoni

Many explorers to Egypt were bitten by the 'bug' and turned into collectors too, and as collecting was lucrative especially with a rich patron it was an attractive side-business. Pietri della Valla (1586–1652) a Roman patrician was in Egypt from November 1615 to March 1616 with the purpose of purchasing antiquities through such an agent or dealer, but also to dig at Saqqara to find his own. Some of the coffins he acquired formed the start of the collection which is known as the Fayoum mummy portraits. His journal entries could be considered the earliest archaeological reports.

However, a particularly lucrative explorer and collector was Giovanni Belzoni (1778–1823), who gave up his career as a circus strongman to travel to Egypt. He was more than six feet tall and part of his strongman act was to lift twelve adult men in a human pyramid. He initially travelled to Egypt to pitch an idea for a hydraulic water wheel apparatus to Mohammed Ali. However, whilst there he could not resist a little bit of sightseeing which included the animals' catacombs and step pyramid at Saqqara and, of course, the pyramids of Giza.

> Though my principal object was not antiquities at that time, I could not restrain myself from going to see the wonder of the world, the pyramids … We went there to sleep, that we might ascent the first pyramid early enough in the morning, to see the rising of the sun; and accordingly we were on top of it long before the dawn of the day. The scene here is majestic and grand, far beyond description; a mist over the plains of Egypt formed a veil, which ascended and descended gradually as the sun rose and unveiled to view that beautiful land, once the site of Memphis.[20]

As his idea of hydraulics was not successful Belzoni became an agent, where his role was not the discovery of monuments for the sake of research but the discovery of portable antiquities which his benefactor, Henry Salt, could sell. This was a major industry in Egypt with benefactors and agents in direct competition with each other to get the best and most lucrative monuments first.

Belzoni tried to prevent the theft of items which he had earmarked for removal, by chiselling his name into the base of them until he was able to transport them. This need for portable items meant that, when he discovered Abu Simbel and cleared the entrance allowing him to enter, he was disappointed as there were not many portable items inside (FIG 8).

When Belzoni arrived at Abu Simbel in 1816 both temples were covered with a huge mound of sand which he started to tackle with forty local workmen. Johann Ludwig Burckhardt (1784–1817) recorded when he visited Abu Simbel in March 1813 that the temple itself had been used as a place of refuge by the people of Ballyane when a Bedouin tribe, the Moggrebyn, had come to the area. He records this group arrived annually and pillaged village after village, with 150 horsemen and the same number of camel riders. The inhabitants of Ballyane headed to the temple with their cattle showing that although the temple was not known by the West at this time it was an important part of the landscape for local tribes. Once in the temple, the villagers were protected and the Moggrebyn were unable to reach them.

It took Belzoni longer than anticipated to clear the temple of Abu Simbel, so he left and returned, only revealing the door in August 1817. He created a space just large enough to wiggle through into the temple beyond. As there were few saleable objects Belzoni made some quick measurements and records, carved his name into the wall and left. Although he had little interest in the architecture and wall scenes, the discovery meant there was a trickle of tourists who then made the trip to the site. However, the site was not 'tourist ready' in the modern sense and the sand was always blowing over the entrance of the temple. Enterprising locals allowed it to become impassable so they could charge visitors to clear it.[21] A century later this was still an issue as James Baikie, a Scottish Theologian (1866–1931) said: 'It is feared that man will, here as elsewhere, wage a fruitless battle in the end against nature, and that the great temple will finally be buried.'[22]

Travelling down to Abu Simbel saw Belzoni and his wife Sarah take in a number of sites along the way including Esna, Edfu, Kom Ombo and Aswan, with Sarah paying a visit to two wives of a local chieftain where she drank coffee and smoked a water pipe.[23] They took an excursion to Philae where they stopped for a picnic, a luxury we cannot imagine doing today. He also earmarked some items he intended to purchase and remove on the journey back, which included a granite obelisk which is currently at Kingston Lacy in Dorset (FIG 9). The journey for this obelisk was not, however, straightforward as it had already be 'claimed' for France so Belzoni basically worked quickly which resulted in it spending some time on the riverbed before being loaded onto a boat to Luxor.[24]

Belzoni's size and his strength was to give him an advantage over other agents. For example, he heard over the grapevine about a fabulous head at the Ramesseum on the west bank at Luxor. Johann Ludwig Burckhardt had been promised the statue known as the Younger Memnon for the prince regent of England (to become George IV), but could not figure out how to transport it back to London. The statue had lain undisturbed for thousands of years alongside a bigger one, commonly referred to as Ozymandias, which Diodorus Siculus described in the first century BCE:

> Beside the entrance [to the temple] are three statues, each carved from a single block of black stone from Syene. One of these, which is seated, is the largest of any in Egypt, its foot alone measuring over seven cubits … it is not merely for its size that this work merits approbation. It is also marvellous because of its artistic quality and excellent because of the nature of its stone, since in a block of so great a size there is not one single crack or blemish to be seen. This inscription on it runs: 'King of Kings, I am Ozymandias. If anyone would know how great I am and where I lie, let him surpass my works.'[25]

The granite statue stood nearly three metres tall and weighed as much as eight tons. The locals were not keen to help move it as they could not understand why the Europeans wanted it – so the Europeans called in the 'Strong Man'. Belzoni started the process in July 1817, just before the Nile was due to flood and make the roads impassable. His plan was to drag the statue across the desert to the river. He raised it onto a sledge using a series of levers, and a team of eighty workmen dragged it across the landscape. This was the same method that had been used in ancient times. They were successful and the Younger Memnon can now be seen in the British Museum. It was this statue which inspired Percy Bysshe Shelley in 1817 to write his famous poem 'Ozymandias', about Ramses II.

Whereas Abu Simbel was thought of by Belzoni as a bit of a flop, the Valley of the Kings in the winter of 1816, on the other hand, was considered a great source of portable artefacts where he discovered the tomb of Ay (WV23 or KV23) and WV25 in the Western Valley,[26] the latter of which contained eight coffins and twenty-second dynasty mummies (945–715 BCE).

Returning to the main Valley in 1817 he discovered the tomb of Montukherkhepshef (KV19) on 9 October, and on the same day an uninscribed tomb (KV21), which housed two female mummies. They were in poor condition and by 'pulling it a little', Belzoni identified that their long hair, 'was easily separated from the head'. This did not alter the market value a great deal. The next day he discovered the tomb of Ramses I (KV16):

> I found the tomb just opened and entered to see how far it was practicable to examine it. Having proceeded through a passage thirty-two feet long and eight feet wide [*9.7 x 2.4 m*], I descended a staircase of twenty-eight feet [*8.5m*], and reached a tolerably large and well-painted room ... The ceiling was in good preservation, but not in the best style.
>
> We found a sarcophagus of granite, with two mummies in it, and in a corner a statue standing erect, six feet six inches high [*1.98 m*], and beautifully cut out of sycamore wood ... The sarcophagus was covered with hieroglyphs merely painted or outlined.[27]

On 16 October he then discovered the tomb of Sety I (KV17). Sety's translucent calcite sarcophagus, minus its lid, was sold by his patron Henry Salt to Sir John Soames for £2,000,[28] and is on display in the Sir John Soames Museum in London. The sarcophagus lid of Ramses III which Belzoni rediscovered in Bruce's Tomb (KV11) was to be sent to the Fitzwilliam Museum in Cambridge. Belzoni cleared Sety's tomb in ten days which to a modern archaeologist is mind-blowing, making some copies of the illustrations, removing some inscriptions and taking squeezes of others before moving on to the next project: the Colossi of Memnon and the associated temple.

Belzoni, like most people visiting Egypt in the eighteenth century, was not trained in archaeology as the discipline did not really exist. However, he was considered an expert as the more monuments and artefacts one person discovered, the more they were appreciated for their skill, and even today he is revered as one of the founders of modern Egyptology. At the time, he was approached by Lord and Lady Belmore when he was in the Valley of the Kings and asked his advice on a good place to start digging, clearly trusting his experience and judgement. He showed them a good spot and they soon discovered two mummy pits which have since been numbered KV30 and KV31.[29] Literally anyone could dig in Egypt if they had the means to travel and to spend time, money and energy on a project.

## Chroniclers

Alongside the treasure hunters were also those with an interest in the history and recording it. Some of the first people to travel with the intention of recording rather than pillaging the monuments were Benoît de Maillet (1656–1738) and Paul Lucas (1664–1737) who decided to carry out an elaborate study of the monuments at Thebes and Jean de Thévenot (1663–67) who claimed he travelled purely for a 'curiosity and a passion for learning'. When visiting Giza and Saqqara he recorded his findings of the pyramids which he believed quite rightly to be nothing but a tomb, 'for the pharaoh who by the will of God was drowned in the Red Sea with all his army'.[30]

One traveller known literally as the Anonymous Venetian wrote an account of his trip to Egypt called, *Viagio che ò fato l'anno 1589 dal Caiero in Ebrin navigando su per el Nilo* ('A journey made in 1589 from Cairo to Ebrin sailing up the Nile') which is currently in the National Library of Florence. He travelled from Cairo almost to the second cataract and chose to sightsee along the way, which in his own words for not for any 'practical purpose': 'For some years I had had the desire to see the province of the Saites, which is the end of the land of Egypt, and so I did, not for any practical purpose, but only in order to see so many superb edifices, churches, statues, colossi, obelisks and columns ....'[31] He visited Karnak temple and marvelled, as many did before and have done since, at the hypostyle hall and the vast columns which were 'shaped like trees' and made him think that he 'must be dreaming'.

In 1707, Jesuit priest, Father Claude Sicard travelled the length of Egypt to Aswan and was the first European to reach that far south. He was to make four visits to Egypt before he died of the Cairo plague in 1726. He had been commissioned by the French regent to make his journeys and to record the monuments that he viewed along the way.[32] He carried a copy of Strabo (63 BCE–23 CE) and Diodorus with him and consulted them in the same way a modern tourist consults the Lonely Planet guide. The records he made were stored in the Bibliothèque Nationale where they were consulted by later visitors. He made various stops on his journey visiting twenty pyramids, twenty-four temples and discovered at least ten tombs in the Valley of the Kings in Luxor.[33] Although he was not an archaeologist, and the discipline of archaeology did not exist in any tangible form at the time, he understood that what he had discovered were tombs:

> These sepulchres of Thebes are tunnelled into the rock and are of astonishing depth. Halls, rooms, all are painted from top to bottom. The variety of colours, which are almost as fresh as the day they were painted, gives an admirable effect. There are as many hieroglyphs as there are animals and objects represented, which makes us suppose that we have there the story of the lives, virtues, acts, combats and victories of the princes who are buried there, but it is impossible for us to decipher them for the present.[34]

British explorer Richard Pococke, wrote a lengthy description of the Valley of the Kings when he visited in 1739, as well as leaving his name carved into the wall of one of the tombs – although this is now lost.

> We came to a part [of the path] that is wider, being a round opening, like an amphitheatre, and ascended by a narrow steep passage about ten feet high [*3 m*], which seems to have been broken down through the rock. By this passage we came to the Bibal el Meluke [*Valley of the Kings*]. There are signs of about eighteen tombs if I made no mistake. However, it is to be remarked that Diodorus says seventeen of them only remained until the time of the Ptolemies: and I found the entrance to about that number, now there are only nine that can be entered into.[35]

He had marked the tombs on a map, and whilst most of them match up to known tombs, there are some that have acted like a treasure map over the years.

When Scotsman James Bruce (1730–1794) visited the Valley of the Kings in 1768, he claimed the tomb of Ramses III (KV11) as his own, although this may have been one already discovered by Sicard. Regardless, it was known locally and in Europe as Bruce's Tomb.

> It [the Valley] is a solitary place: and my guides, either from a natural impatience, and distaste that these people have at such employments, or that their fears of the banditti that live in the caverns of the mountains were real, importuned me to return to the boat, even before I had

begun my search, or got into the mountains where there are many large apartments of which I was in quest.

    Within one of these sepulchres, on a panel, were several musical instruments strewed upon the ground, chiefly of the hautboy [oboe] kind, with a mouthpiece of reed … in the three following panels were painted, in fresco, three harps, which merited the utmost attention … With great clamour and marks of discontent … dashed their torches against the largest harp and made the best of their way out of the cave, leaving me and my people in the dark: and all the way as they went, they made dreadful denunciations of tragical events that were immediately to follow upon their departure from the cave. There was no possibility of doing more.[36]

Tourism and adventuring at the time were a lot more dangerous than in the nineteenth century and he records that when he was drawing some of the harpist scenes from the tomb walls, he was forced to leave due to the threat of local bandits,[37] which is a lot more hair-raising than the modern experience of having to leave to go to the perfume factory. The harpist scenes were the first tomb images to be published in 1790 and completely fascinated those who saw them.[38]

## Dangers of Travel

Vitaliano Donati also records some of the dangers of his trip in 1759–60. He started his trip in Alexandria and went south to the first cataract in Nubia. He had reached the village of Neg (el) Heseitam or Neg (el) Heseilam according to his diary, which may be to the north of Nubia, but was advised not to go any further. Sheikh Amman, the local ruler, explained there were groups of bandits in the area who were not under his jurisdiction and that it would not be safe to continue his journey. Donati was forced to abandon his planned excursion of Nubia and return north to Cairo.[39] However, on the return journey when he reached Manfalut, just north of Asyut in middle Egypt, he had an altercation with some thieves, showing the problem was country-wide and with no established tourist trails and accommodations there was no official protection for travellers. Therefore many tourists carried a letter of introduction (a.k.a. a firman) which enabled them to travel safely along the Nile, under the protection, as it were, of a type of benefactor. An example of such a letter (dated 1737) read:

To Emir Mahomet Kamali,
What I order,
The person that brings this letter is an Englishman, going into Upper Egypt, to whatever is curious there; so when he delivers this letter take care to protect him from all harm; and I command you again to take care of him. I desire you not to fail of it, for the love you bear us.
Osman Bey Merlue.[40]

Such a letter was originally written in Arabic and then shown to certain individuals along the way, to expedite arrangements and ensure safe passage. The traveller had to negotiate every aspect of the trip on the ground, using local rais (boat captains) and dragomen (guides). However, even the letter could not protect them from bandits and boats themselves were often the target of attack where:

> a certain kind of robber, peculiar to the Nile is constantly on the watch to rob boats, in which they suppose the crew are off their guard. They generally approach the boat when it is calm, wither swimming under water, or, when it is dark, upon goat skins, after which they mount with the utmost silence, and take away whatever they can lay their hands on. … The attempts are generally made when you are at anchor, or under weigh, at night, in very moderate weather, but oftenest when you are falling down the stream without masts.[41]

Such dangers were not uncommon and there are many tales of being attacked such as that of Jean de Thévenot who, when on his way back to France following his trip (1663–67) was captured by pirates and then rescued by an English warship. Another traveller, the Czech composer Kryštof Harant (1564–1621) recorded in 1598 that when visiting the Sinai he was attacked by bandits who stole most of his possessions.

When on the Nile the dangers could be more practical as the Nile itself was not the landscaped and engineered river we see today. When Fredrich Louis Norden (1708–1742) reached Luxor he was very excited at the prospect of getting off the boat in order to see the ruins of Luxor and Karnak temple, but his rais refused to stop the boat because of 'the impossibility of landing, on account of the islands and sandbanks'.[42] This may of course have been an excuse, but the plausibility of the difficulties indicates that there were no engineered landing spots at the sites and the boats had to stop anywhere along the bank and hope they would not get stuck on a sandbank.

## An Early Itinerary

Vitaliano Donati travelled to Egypt in 1759–62 and kept extensive diaries about his trip providing invaluable information about travel in the eighteenth century. His journey followed a familiar pattern starting at Alexandria at the end of 1759 to the beginning of 1760. In his time there he visited Pompey's Pillar and the catacombs of Kom El Shoqafa. Then he travelled to Rosetta, just north of Alexandria and where the Rosetta Stone was found some decades later. He arrived there on 24 February 1760 and stayed for five days.

Whilst in Rosetta he was more interested in the local villages and customs and recorded the landscape and local practices. From Rosetta he passed 127 towns and villages, before arriving on 9 March in Cairo where he spent four months visiting the pyramids at Giza, along with Venetian consul Bernadino Ferro, Stefano Aspahan the translator and a small entourage. Here Donati made some drawings of the pyramids, but the wind and the sand made it problematic, and he was not pleased with the results. He collected a number of rocks and pieces of 'cement' from the pyramid to take back as a research tool.[43] On his return journey to Cairo in December of that year, he described visiting the Convent of Saint George (the Babylon Fortress in Coptic Cairo) and the Monastery of the Martyrs. Leaving Cairo, he arrived at El Minya in middle Egypt in July 1760, but did not leave the boat to visit the town as he was warned it was not safe to disembark.

He then made some whistle-stop visits to Asyut, Abu Tig and Tahta between 13–15 July, where at the latter he visited 'serpent mountain' where the locals worshipped a large snake. He then arrived at Akhmim on 16 July. This was a very popular town to visit in the Roman period (chapter 6), but it had fallen off the beaten track by the nineteenth century entirely. When Donati was there he records visiting churches, but no ancient remains. On his return journey he stopped here again on 10 December, where he discovered a large stone bearing a Greek inscription.

He continued his trip up the Nile, stopping at Girga, Dendera, Qena, Koptos, Naqada, Armant, Esna, Redesieh and Aswan, reaching Sudan in Nubia on 5 September 1760, where he stopped at the first cataract. He then started his journey back to Cairo where he visited the more traditional ancient sites that still appear on the standard tourist itinerary arriving in Luxor on 11 September. He stayed for three days and visited Medinet Habu on the west bank along with the Colossi of Memnon.

When he arrived at the Valley of the Kings, the Nile was in inundation and he described that in order to reach the monuments they used temporary boats made from maize stalks, as well as being carried on the shoulders of

Egyptian sailors. He thought Qurna (Valley of the Nobles) was inhabited by thieves and described how many of the tombs (caves) had been turned into houses by the locals. He purchased a number of mummies whilst he was here, but clearly did not associate purchasing antiquities in this way as theft or morally questionable.

Returning to the east bank he stayed at Karnak 'working from sunrise until midday without interruption' on recording the temple ruins giving some indication of the tranquillity of the site at the time. Whilst here, Donati also did a little excavation work uncovering two statues of Ramses II and Sekhmet which he thought were Osiris and Isis.[44] When he reached Mallawi, just north of Asyut in middle Egypt, he was required to pay a 'tribute' to a local leader to keep the antiquities he had collected on his journey. He arrived back in Cairo on Christmas Eve 1760.

Donati travelled as part of a small group, which was normal for anyone travelling for leisure, as an agent or for general explorers. It was safer and may have reduced some of the dangers of travelling. However, there were, of course other benefits. Danish explorer, Frederick Norden (1737), was helped in his voyages by Christian VI, the King of Denmark and wrote *Voyage de l'Egypte et de la Nubie* where he commented that the best way to view the pyramids was 'with a party; they mutually excite each other's curiosity'.[45]

The journey from Cairo to the Giza pyramids took two days by donkey and travellers likely stayed overnight as Norden explained: 'There a very disagreeable night is passed by the curious, without beds, or other convenience, they are tormented by bugs, but one night is soon over, and when curiosity eggs, such difficulties are most easily born.' This is very different from the tombs set up with beds and others as restaurants that had evolved in the nineteenth century (chapter 2).

When the group arrived at the pyramid, they 'fire some pistols', to get rid of any resident bats and the Egyptian guides removed the sand from the doorway before entering the sweltering monument. Every visitor held a candle to light their way through the structure, and the corridors were very narrow:

And through this hole must curiosity pass, on belly couchant, while the two Arabs who have wriggled themselves through before, seize each leg, and drag their gentlemen through this probation cleft, all covered with filth. Happily, this narrow passage is not about two yards long … then a large space opens, where the traveller takes breath, and some refreshments.

When they left the pyramid in the relatively cool air of the desert plateau, they drank a medicinal glass of wine before climbing to the top of the pyramid to take in the views. When John Lloyd Stephens, an American explorer, visited the pyramids in 1836 he was in awe of the size and commented 'how very small I was', and as he glanced up at the summit 'they seemed to have grown to the size of mountains'.[46]

It seems that for time immemorial, when tourists have seen tall monuments, regardless of what they are, the automatic response was to try to climb them. Dr R.R. Madden in 1825 talks about visiting Pompey's Pillar in Alexandria, and his attempts to climb the nearly 27-metre-high monument:

> I have seen ladders rigged to the top by English sailors, who contrived to pass a rope over it by a common kite. I made two fruitless attempts to ascend, but I found it impossible; an Irish lady, however, a Miss Talbot, had the courage to mount, and breakfasted on the summit.[47]

Even travelling back in time by a century (to the seventeenth) shows how far the tourist trade had evolved. Travelling at this time was rough and ready with few comforts of home, offering the 'real' adventure of a trip to a vastly unexplored (at least by Westerners) country. However, the collectors, antiquarians and explorers of the seventeenth and eighteenth century were already walking in the footsteps of thousands of visitors before them.

# Chapter 4

# Christian Pilgrimages to Egypt

During the period between the Muslim invasion in the seventh century and the seventeenth century, the main draw for European travellers to Egypt was Biblical associations. During the fifteenth and sixteenth centuries there was more widespread accessibility to the Bible which led to an increased interest in Egypt and the Middle East. The Bible discusses Egypt in various chapters, meaning European Christians were familiar with the country and the importance it held to Christianity. Although a trip to the Holy Land and Jerusalem was the main goal of Christian pilgrims, Egypt became a popular stop along the way due to the notion that: 'Ancient Egypt somehow stands at a point of origin for Western civilisation, has its roots in Biblical and classical tradition.'[1]

The key factor at this time, was that the Bible was taken as fact and therefore an important element of Middle Eastern history. With this in mind, James Usher, Archbishop of Armagh, in 1650, calculated that the world was created on 23 October 4004 BCE, at noon. He was able to show his workings which were based in Biblical teachings, along with his assumption that the world would only last for 6,000 years. Using the death of Herod in 4 BCE as a starting point he came to the conclusion that Jesus must have been born in the same year.

Usher chose noon of the first Sunday after the autumnal equinox as the creation time, as day and night are equal in length.[2] This meant that to many Renaissance Europeans, pre-history did not exist, and everything had to be crammed into a short timeframe. This became an amazing feat of historical Tetris, as they tried to shoe-horn the reality of pre-history, geology, and science into these tight boundaries and apply it to the classical culture they were able to witness in Egypt and across the Holy Land.

This struggle continued until 1859 when Charles Darwin rocked the boat with his *On the Origin of Species*, which introduced the concept of a much longer world history bringing the wrath of the Church upon him.

## Christian Pilgrims

Egypt was fascinating to Christian pilgrims and there were a number of important pilgrimage sites to visit, including the shrine of Apa Mina, to the west of Alexandria, the oracle of Saint Philoxenus in Oxyrhynchus, the healing cult of Saint Colluthus in Antinoë, and Askun, the site where Moses was set adrift in his basket by his mother. However, many of the sites were not enticing enough on their own as they lacked direct Biblical associations, meaning the cult of sacred relics was an important development in attracting pilgrims.

Throughout the height of the Christian pilgrimages, the motivation was very much one of spiritual growth.

Tourism as an activity in itself does not seem to be as prevalent as in earlier periods. However, unlike other periods of history, the archaeological record of pilgrims and travellers is lacking, and it is essential to rely on religious texts as well as books of pilgrim itineraries such as Palladius's *Historia Lausiaca* (fifth century) and the Historia *Monachorum in Aegypto* (fourth century). Unfortunately, as many of the sites are no longer standing, travellers' graffiti has also gone, meaning for Christian travel we are reliant on religious records produced by the Church which were primarily written to promote the power of God, but also used to bring more visitors to a particular site. Whilst some records mention pilgrims by name, there is some doubt as to whether many were real people rather than literary tools. Even if they did exist and visited religious sites, the records may have been curated to promote the intended message. However, such issues aside, the surviving records are useful as they do explain the proposed motivations of visitors.

Accordingly, common people went on pilgrimage to be cured of illnesses, for the future health of a child, for baptism, to address a particular issue, for an easy birth, to find lost objects or to discover whether someone was telling the truth.[3] The more elite pilgrims may have visited sites to speak to clergy for advice on matters of state or scripture. Essentially, Christian pilgrims were using these sites in the same way as pharaonic Egyptians had used oracles for centuries (chapter 7), showing a continuation of cultural practices.

Evidence also suggests that some travellers visited a particular church, shrine or hermit to have a priest carry out an exorcism. This was seen as a speciality of the localised Egyptian form of Christianity and there was a performative element to the ritual. The priest Athanasius of Alexandria (d. 373 CE) records witnessing exorcisms at the shrines of martyrs which acted as a tool to convert non-believers who stood in awe of the clergy's powers.[4]

Only having records from this type of resource presents a focused, and not particularly human, approach to travel. If they are to be believed, travellers only went to religious sites, bypassing the pyramids and the Sphinx, Karnak Temple and the Valley of Kings in favour of monasteries and churches. Whilst this may have been the case for some travellers, it is likely that those who saw them wondered at the size of the pyramids, or marvelled at the paintings in the tombs in the south.

It is clear from records, however, that during the European Renaissance (fourteenth to seventeenth centuries) searching for knowledge and truth became a *raison d'être*. As the end of this era was a century or more before hieroglyphs were to be deciphered, many scholars turned to Greek and Roman texts, as well as the Bible as resource material. Here it is stated that the ancient Egyptians held the secret to great knowledge: 'And Solomon's wisdom excelled the wisdom of all the children of the east country, and the wisdom of Egypt. For he was wiser than all men.' (1 Kings 4:30–31) The Old Testament also alludes to this vast knowledge: 'And Moses was learned in all the wisdom of the Egyptians and was mighty in words and in deeds.' (Acts 7:22).

Moses was said to have written the Pentateuch, the first five books of the Old Testament, the lives of the Patriarchs and the account of Hebrew slavery in Egypt. To many European or Christian travellers this raised the question as to whether the Egyptians learned their wisdom from the Jews, or the Jews learned it from the Egyptians. Theologians from Clement of Alexandria (150–215 CE) to Theophilus Gale (1669) believed the Egyptians learned from the Jews when Joseph was exiled in Egypt, whereas the Hebrew scholar from Cambridge John Spencer (1685) believed the opposite to be true.[5]

A revival in Neoplatonism and Platonism further connected Egypt in the mind of these scholars with wisdom and knowledge.[6] This therefore meant that many scholars and theologians travelled to Egypt hoping to learn something of ancient knowledge from the priests who were the final link to the ancient Egyptian people. One such traveller was Ciriaco of Ancona (1390–1452 CE) who spent nine days travelling in 1434 from Alexandria to Cairo, where he commented on an inscription he saw: 'I think it is unknown to the men of our era on account the antiquity and the ignorance and disuse of the great and ancient arts.'[7] This perception did not prevent him from copying the inscription and passing it on to Niccolò Niccoli 'a man most interested in this sort of thing', to try to decipher.

## The Journey of the Pilgrim

Towards the end of the Roman period pilgrimages to churches and monasteries in the north of Egypt were popular as were those to the Eastern Desert and Sinai Peninsula. Some remain important sites to the modern day. A travellers' itinerary to Egypt included places mentioned in the Bible or in the classical sources such as: Ramesse, Migdol, Etham, Tanis, Abydos, Philae (Fialus), Memphis, Babylon, Pelisium, Alexandria and the Lighthouse of Pharos.[8]

Many of the churches popular with pilgrims were regularly looted in times of political instability for the wealth held within, meaning they were often demolished, restored and rebuilt many times over the centuries. There was also an interesting conflict of interest with monasteries which attracted pilgrims as a form of income. They were often remote to afford the monks solitude, which of course was interrupted by travellers and pilgrims.[9] However, some pilgrims stayed for an extended period of time becoming monks themselves and therefore expanding the monastery.[10]

## The Flight of the Holy Family

The Flight of the Holy Family into Egypt laid the route for many pilgrims to Egypt. Mary, Joseph and the infant Jesus stayed in Egypt for three and a half years, 1,260 days in total, and in the eleventh-century *History of the Patriarchs of the Egyptian Church* there were ten sites on the itinerary mapping their journey. In the Coptic *Psali*, there are only six sites and currently the route sits at about thirty sites. It is thought that some bishops had, over the centuries, extended the journey to incorporate their dioceses.[11] By creating an itinerary of the holy journey, the Church was able to connect the Gospel and the landscape of Egypt by making ritual connections between the two.[12]

Many of the sites on the journey were where the infant Jesus produced a well, a rare and important resource in a desert environment and included: Musturud, Tell Basta, Antinoë, as well as a site eight kilometres from Meir where Salome bathed Jesus and Mary gave him milk, Sakha, near Kafr al-Shaikh in the Delta, where Jesus stepped on a rock from which a well formed, leaving a footprint in the rock, Harat Zuwayla where the infant Jesus blessed the well and El Ashmunein where Jesus is said to have raised a person from the dead.

Many of the stops on the route were not large churches with relics but areas of nature, such as the tree of Matariya, in a suburb of Cairo. Wherever the Holy Family had stopped to take shade became an important pilgrimage site and this was one of the largest cults during the Middle Ages and one which continues to the modern day. The tree of Matariya was believed to have offered shade to the Holy Family and the *Ethiopic synaxarion* records the event:

> Then said Jesus unto his mother: 'We will tarry here' and that place and its desert and the well became known as Matariya. And Jesus took

Joseph's staff and broke it into little pieces, and planted those pieces in that place, and he dug with his own divine hands a well, and there flowed from it sweet water, which had an exceedingly sweet odour. And Jesus took some of the water in his hands and watered therewith the pieces of wood which he had planted, and straightaway they took root, and put forth leaves, and an exceedingly sweet perfume was emitted by them. And these pieces of wood grew and increased, and they called them balsam. And Jesus said unto his mother, 'Oh my mother, these balsam which I have planted shall abide here for ever, and from them shall be taken the oil for Christian baptism.'[13]

The earliest written text recording this is the *Arabic Infancy Gospel* dated to the sixth century where Mary is also recorded as washing the infant Jesus' swaddling clothes and hanging them on a post to dry.

Thirteenth-century pilgrim Burchard of Mount Sion visited the site and took away some sticks from the balsam trees and bathed in the water where Mary had dipped Jesus. The balsam in the fourteenth century was said to have healing properties with Muslims believing it was good for nasal issues, lumbago and knee pain, while Christians believed it was ideal for toothaches, and for curing poison and snake bites.

In 1435, according to Pero Tafur, only five pilgrims at a time were allowed to visit the tree of Matariya and they were monitored to prevent them from removing any branches or leaves. They were charged six ducats to enter the garden, where they could bath in the pool and enjoy the healing environment.

The site was extended by 1483 when Dominican Friar Felix Fabri, recorded a fig tree near the entrance to the garden which was hollow and contained a chapel, as the tree had apparently provided refuse to the family when they were being chased by bandits.[14]

The amir Yashbak built a domed house in this area which was viewed as a form of paradise, and he entertained Qaitbay (1467–96 CE) here. By the end of the fifteenth century a chapel had also been built containing a pool of water for healing pilgrims and a niche which emitted a sweet smell where Mary had once placed her son.

However, when Qaitbay died, the knight Arnold von Harff records how the balsam trees were pulled up and the garden with the water wheel was destroyed. The new Sultan, Salim I (1517–20) wanted balsam trees from the Hizaz region to be planted in the garden instead. The new trees only survived for a few years. In 1575, the Turkish Pasha tried to plant balsam trees from Mecca, but they had also failed to grow. Between 1638 and 1645 Father Jean Coppin, the French consul, had only heard about the balsam trees as they were now long gone from the garden. The remaining sycamore tree at the site soon became the focus of attention for pilgrims, both Christian and Muslim. However, the tree was not in good condition, and it had to be replanted in 1672 and survived until 1906.

Another important site was the Church of Saints Sergius and Bacchus which was built over a site where the Holy Family were believed to have taken refuge, and the crypt was dedicated to this event. The church still stands in Coptic Cairo. On the first day of June (Day 24 of Coptic month Bachons) medieval pilgrims celebrated the Flight into Egypt here with a mass. The church itself is dedicated to Sergius and Bacchus who were martyrs during the reign of Roman Emperor Maximinus (173–238 CE).

Another church on the route of the Flight of the Holy Family was the Church of the Virgin Mary in Babylon Al-Darag, although this association was not made until the time of Patriarch Cyril II (1078–92 CE). It was also the burial place of the Patriarch Zacharia (1004–32 CE).

## Stational Pilgrimage

As well as following the route of the Flight of the Holy Family, some shrines had also introduced the idea of a stational pilgrimage, where there was a specific route around a shrine with prescribed hymns, prayers or actions at each station. The idea of movement was seen by some as a means of tracing holiness by following a particular path or route and for others continual wandering was seen as their path to holiness. There were therefore numerous aesthetics, both men and women who wandered the desert in order to find God.[15]

This idea of movement had been in place since at least the Roman period when an inscription at the temple of Mandulis at Kalabsha states: 'Revere the divine. Do ritual honours to all the gods. Travel to make worship at every shrine.'[16]

The Feast of Apa Shenoute, which was held at Easter at the White Monastery included a procession where there were specific rituals to be carried out at particular locations. When Shenoute was alive, the White Monastery was an important destination for Christians. It first appears in the records in 743 CE, but is likely to be much older, as old as the fourth century and located near Sohag, the ancient city of Panopolis. Shenoute was an Abbot (d. 466) and many pilgrims thought he possessed supernatural powers and visited his monastery for spiritual and intellectual nourishment. There was originally a library here of his works which also attracted theologians and scholars.

At its height the monastery housed thousands of devotees, both male and female and in *The Lives* written by Shenoute it mentions there were many reasons to make pilgrimage here: to make an offering (gift), to receive a blessing, peace of mind, or to receive absolution. He even records one pilgrim who was a robber and murderer and was hoping to receive forgiveness.[17] Shenoute wanted to promote the monastery as an alternative to the Holy Land:

> He who cannot visit Jerusalem in order to prostate himself before the cross on which Jesus the Messiah has died should come to offer in this church together with all the angels, and I shall pray for the sins they have committed previously, and whoever hears me, his sins shall not be held against him, even including the dead buried in this mountain, because I shall intercede with the Lord on their behalf.[18]

Shenoute held public sermons which were very popular with local pilgrims, both officials and common people. In one of his records he mentions that on one occasion 'I will cut my sermon short and stop, for I see that you want to be going,' indicating he took notice of the people who came to hear his wisdom. Following the Saturday service there was also a distribution of bread to the poor.

The records show that many officials made the journey here, including provincial governors from the Thebaid, as well as from Lower Egypt, in order to ask for advice on the scripture and moral conduct. In one instance a governor came to the temple to get Shenoute's blessing on a battle that was being staged against local tribes. Shenoute gave the governor a wooden cross he carved himself to wear as a protective amulet.[19] Roman government officials treated Shenoute with respect as a secular power in the area, and therefore someone to keep on side. The White Monastery also acted as a haven for refugees from Upper Egypt who were threatened by the Blemmyes tribes (a desert nomadic tribe). These tribes themselves became a target for Shenoute's mission, in line with the government policy against pagans.[20]

Other people visited the monastery as a way of settling disputes such as thefts and were to a certain extent treating the monastery in a similar way that Greeks and Egyptians used oracles (see chapter 7). The records show one visitor came from Komentios (between Koptos and the Red Sea port of Berenike) and another from Oxyrhynchus/Behnasa with the intention of donating 120 gold coins to the monastery.[21] It is likely that many officials travelled from Alexandria to the monastery as needed. However, there is no surviving evidence demonstrating there were foreign pilgrims to the site.

One ritual procession is recorded in a manuscript in Paris (FR-BN Copte 68) which dates from the fifteenth to sixteenth centuries, rather than from records of the travellers themselves. It appears to have been copied from earlier texts as this procession would not have been possible at this time due to the disrepair of the site. The ritual was carried out from Shenoute's death in 464/465 CE until approximately 1200 CE and the route gives us an idea of the landscape at the time before the monastery and the estate fell into disrepair. The estate attached to the monastery covered a few square miles and included a number of out buildings and structures. Most of these buildings were completely ruined by 1450 CE with only the main church standing.

This particular stational procession was held at the beginning of week two of Lent, before Holy Week and Easter. For the procession the pilgrims, many of whom came from Akhmim followed a strict route and performed rituals at each station such as prayers, hymns or Eucharist:

- Pilgrims gathered at the 'corner of the choir master', indicating that the procession begun with singing of Biblical passages led by the choir master.
- They then went up the 'mountain' which is likely to be the Atripe Mountain or may have just been an elevated area in the desert meaning pilgrims from Akhmim needed to 'go up' to reach the site.
- Then they headed north to the church. This is a church on higher ground than the monastery itself and known in Arabic resources as er-Rogamah. It is unclear which saint this church was dedicated to, although it is thought it may have been the Virgin Mary.
- The pilgrims then went to a 'holy place' where they listened to one of Shenoute's sermons discussing sin and the appropriate punishments.
- The pilgrims then processed to the 'sea' whilst reciting psalms and prayers which reference the sea or water. The sea is believed to be a cavity beneath the altar which held relics (bones), of Shenoute. This is likely to be in the Church of the Virgin Mary.
- This was followed by a procession around the church before mass was celebrated.
- They entered the monastery of Apa Shenoute stopping at the door of the monastery and the door of the church of the Virgin Mary and Saint George. The pilgrims sang 'Greek hymns (psali) for the feast of the Lord and the feast of the desert.'
- They visited both the church of the Virgin Mary and of Saint George where the pilgrims then requested healing should they require it.[22]

Unfortunately, the manuscript describing the route has ten pages missing from the point they reached the mountain. These pages may simply contain the passages and psalms to be recited or they could include further sites before the procession headed to the church. The end of the procession is also rather hazy but provides an idea of what a pilgrimage to this monastery and surrounding site was like.

## The Cult of Relics

One of the key aspects of the procession was coming into contact or close proximity to the relics of Shenoute, which were the focus of the main church. This was an important aspect for all pilgrimages in Egypt. Many of the important Christian pilgrimage sites were based upon the cult of sacred relics which were often in the form of skeletal remains. Although this was an important part of worship not everyone approved – in the fourth century Athanasius of Alexandria (d. 373 CE) decried the worship of martyr's bones, the invention of holy places and the devotion he saw at these sites.[23] Despite such opposition, relics were an important 'marketing tool' and enabled emperors and church leaders to raise the profile and importance of a church, site or city by introducing relics to it.

However, not all relics were created equal and Biblical saints were the real prize and could raise the profile of the church or city by bridging the gap between Christian Egypt and the Holy Land. Unfortunately for many sites, Biblical relics were not available and instead they housed the bones of Egyptian saints and martyrs which in themselves acted as mediators between local Christian belief and the wider Church ideology.

In the fifth century Archbishop Theophilus (d. 412 CE) endeavoured to associate each of his new Christian cities with a saint or martyr. Three of note were associated with Alexandria and were the relics of John the Baptist, Elijah and the Three Hebrews from the Book of Daniel (3:1–30). The shrine to the latter no longer exists but was thought to be located just outside Alexandria.

The Three Hebrews (Ananias, Azarias, Misael) were proto-martyrs under the reign of Nebuchadnezzar II of the neo-Babylonian empire (605/604–562 BCE) and were thrown into the furnace for refusing to worship the king's golden statue. They were unharmed by the fire as they were protected by the power of God.

> And the satraps, administrators, governors, and the king's counsellors gathered together, and they saw these men on whose bodies the fire had no power; the hair of their head was not singed nor were their garments affected, and the smell of fire was not on them. (Daniel 3:27).

However, as with many of the relics that still exist today, even in the fourth century there was doubt as to the location of the authentic bones of the Three Hebrews which had been discovered in 420 CE (at the same time as Theophilus) in Ctesiphon in Babylonia and had, it was believed, remained there.

Such details did not pose a problem for Theophilus, who sent John the Little to Babylonia to appeal to the spirits of the Three Hebrews directly:

> he began to pray, and immediately at that hour a cloud carried him and placed him in Babylon where lay the bodies of the holy and excellent athletes of Christ God. Our holy father saw the perfect gift of the tomb of the saints by means of the light of the Holy Spirit who guided him. He worshipped upon the earth three times before he reached them. When our father approached them he threw himself down on his face and embraced their holy relics with a display of many sweet tears.[24]

John the Little asked the Hebrews if they could transfer their bodies to Alexandria, to which they refused but offered an alternative:

> Let them decorate the shrine [in Alexandria] and hang all the lamps without oil or wicks and gather there with all the people and clergy. At night we will come and place in it the power and blessing of the Lord and in this way sanctify the House of God, ... And by the power of the God of Israel, which the archbishop lives we will spiritually dwell in it through signs and wonders.[25]

John the Little relayed the message, gathered everyone as instructed and, 'in the middle of the night a great light suddenly appeared in the holy place and clouds of pure sweet fragrance were in the air, especially above the city of Alexandria and the dwelling place of the saints'.[26] Therefore, from that day forward, even though this shrine had no physical relics, the saints themselves had sanctified the site and that was just as powerful.

The martyrion to John the Baptist and Elijah the prophet was a centre of Christianity in Alexandria during the fifth century. The Church of John the Baptist was built on the remains of the Serapeum which fell into disuse in 391 CE, and it was important to sanctify the church by getting relics to fill it. It is said that at the martyrion the relics of John the Baptist had been rescued from the emperor Julian (331–363 CE). So whilst there was no genuine connection between the church in particular, or Alexandria in general, and John the Baptist, the connection was enough to make the church a special place and a draw for pilgrims.

Later relics of Elijah were also brought to the church. However, in order for the church to be acceptable to the local Christians it also needed to be connected to an Egyptian saint or martyr, which manifested itself in the form of Macarius a fifth century holy man and bishop from Upper Egypt. When he was brought to the martyrion by his disciples for burial, an Alexandrian archbishop referred to him as an 'unclean Egyptian', and was immediately struck dead by lightning as a mute child announced that John the Baptist and Elijah welcomed Macarius into the martyrion.[27] This addition firmly authenticated the shrine in the eyes of local and international pilgrims.

Many visitors to these churches and monasteries did so to gain comfort or advice, and on the monuments still standing we are able to see the graffiti they left behind, recording their trip, cure or miracle they may have encountered. Although many of the records that do survive paint a very sedate picture of the pilgrimages, the Archimandrite Shenoute (333–451 CE), of the White Monastery presents a different scenario:

> To go to the shrine of the martyr, to pray, to read, to sing psalms, to be sanctified, to partake of the Eucharist in the fear of Christ is well and good … But to talk, to eat and to drink, to frolic, or rather, shall I say to fornicate, and to commit murder as a result of drinking and lewdness and brawls, with complete stupidity, that is lawlessness. While some are indoors, singing, reading, taking communion, others outside fill the place with the din of trumpets and pipes … ye have made the house of God a place to sell honey in, and bracelets. Ye have made the shrines into prowling grounds for your cattle, racetracks for your donkeys and your horses.[28]

This creates a dynamic and colourful impression of what pilgrimages may have looked like – although it should be considered that Shenoute was exaggerating due to his irritation. However, it does demonstrate there were two acts during a pilgrimage. The pious acts of worship and the celebratory acts afterwards. Many pilgrimages lasted between three and seven days and started on the day before the vigil which could be celebrated with a procession where banners and relics were paraded through the streets and ended after the Divine Liturgy.

Although the sites were Christian, visitors also included Muslims, and theologian Ibn Taymia records that in the thirteenth century Muslims were emulating Christian baptism whilst on pilgrimage, especially in Upper Egypt. The elite were more likely to go on pilgrimage for 'public display but also to do penance for their greater and more public sins.'[29]

## Important Churches

A short distance from the Church of Sergius on the journey of the Holy Family was the Church of Saint Barbara which was reconstructed in 1072–73 CE in order to receive the saint's relics which are still in the church to this day.

It had originally been built by Athansius, a secretary of Abdel-Aziz Ibn Marwan, Governor of Egypt (685–705 BCE). Saint Barbara was thought to have lived in the third or fourth century, and to protect her from conversion to Christianity, her father built a tower to house her in. Despite this, she converted to Christianity and her father handed her to the Roman prefect Marcian who tortured her to convince her to denounce Christ. She refused and Marcian decapitated her. He was struck by lightning and killed on his way home.

### The Hanging Church (Al-Mo'Allaqa)

Another popular church for Christian pilgrims was the Hanging Church (FIG 10) which was dedicated to the Virgin Mary. It was constructed over the gate of the Babylon fortress and was first mentioned during the reign of Patriarch Joseph (831–849 CE) when it was partially destroyed on the orders of the governor of Egypt, indicating it was much older than this date.[30] The church is still standing and is a popular destination for tourists and Christian pilgrims to Cairo.

### The Church of Saints Cyrus and John

The Church of Saints Cyrus and John was one of two major pilgrimage sites in late antiquity although the church seems to have fallen into disuse in the ninth century. In 391 CE, the relics of Cyrus and John, two physicians were transferred from the Serapeum in Canopus to the church of Saint Mark in Menouthis (Abuqir). Cyrus was a doctor or monk, and John was a soldier. Both were Egyptian martyrs during the persecution of Diocletian (303 CE). The only surviving records from the church are written in Greek in the works of Sephronius, the patriarch of Jerusalem. The visitors recorded here appear to be from all social classes from the poor to important officials like Christodorus, the oikonomos (household manager) and his wife and daughter, and another traveller, Ammonius.

Menouthis had been an important cult centre for Isis with an oracle and healing cult and the area continued to be a centre of healing. Visitors came to the church for incubation in the presence of the relics of the martyrs as a form of healing. For many, coming here was a final attempt after visiting doctors and having no respite. Here there were a series of baths, wells and fountains within which sick pilgrims could bathe in the sacred waters but also an area for incubation, where pilgrims slept and received dreams which explained what was needed to instigate their cure. However, for many simply being at the site was enough. For example, Theopompus, was too exhausted when he arrived at the site to make it into the church, so slept outside. He received a dream from the martyrs showing that he did not need to be in the church to benefit. Often the church was so overcrowded there was no room for any more pilgrims inside, so this miracle was useful to show that proximity was key.

There were two types of dream received through incubation, the first was like Theopompus' where he dreamt of the martyrs and was cured straight away, whereas the second was where there were instructions in the dream for a remedy which could be realised upon waking.

Another pilgrim, Jon the sub-deacon from Cynopolis in Middle Egypt, was unable to enter the church because it was full and had to sleep 'where all the pilgrims stay who cannot find another place because of the crowd of sick who are there'.[31] However, cash was able to buy a space in the church, and in the miracle of two ladies called Juliana, one was rich and paid to sleep near the relics, and the other was poor and slept outside the church. Poor Juliana was cured first to show the benevolence of the martyrs.

Some incumbents did not receive a dream and instead received a vision. For example, a paralysed man from Constantinople, Zosimus prayed at the church for three days before being taken to the bath. It was here he received the vision of Saint Cyrus 'not in a dream, as he appears to most people, but in a waking vision, and just in the manner that he is represented'.[32]

Sophronius records that he was cured of an eye complaint by Saints Cyrus and John and adds it to his list of seventy miracles at the site. He divides them into place, which is useful for our purposes here and records thirty-five from Alexandria, fifteen Egyptians and Libyans and twenty foreigners showing the diversity of the reach of the church. It must be considered that there was an agenda with this text and knowing if these people existed, let alone whether they were cured will remain a mystery. The pilgrims from Egypt were generally within easy travelling distance to the shrine, including Nikiou, west of Alexandria, Heraclius (2.5 kilometres from Menouthis), from the Delta towns of Pelusium and Sais, a child from Babylon (Cairo) and three from Mareotis.[33]

People were encouraged to show their gratitude for the cures they received which was in the form of money or offerings and two women showed their gratitude with some money and a pig. Unfortunately, the person asked to deliver the offering stole it and ran off but was stricken down with an illness in traditional Biblical style. Nemesion, a former prefect, donated to the church in the form of marble decoration near the shrine which may have been a mosaic. The images showed Christ, John the Baptist, Cyrus and the pilgrim and offered an insight into how the church may have looked.

Even poorer offerings were welcomed, and John from Rome who was blind came to the church in hope of a cure. He was cured after eight years incubation. He leapt from his bed and recorded the miracle on the wall by the church door, 'I John, from the city of Rome, a blind man, waited eight years here faithfully and recovered my sight through the power of Cyrus and John.'[34]

During the ninth century, there were so many miraculous stories about pilgrims being healed, it was one of the most popular resorts at the time.[35] It is a perfect example of how the traditional practices of earlier Egyptian religion (see chapter 7), had been adopted by a local Christian cult in order to ease the transition to a new religion.[36] Visitors to this cult centre were both local and international.[37] However, this idea of adopting older practices creating a localised Christianity was not appreciated by everyone and Shenoute commented:

> In the moments of suffering … when they fall into poverty or become ill – or indeed other temptations – abandon God and have recourse to enchanters or oracles or … or deceptive things; just as I myself have seen – the snake's head bound to the hand of some, and another with the crocodile tooth bound to an arm, another with fox claws bound to his legs … I reproachfully asked him whether it was the fox claws that would heal him, he said 'It was a great monk who gave me them saying, Bind them to you, and you will recover.'[38]

Unfortunately, not everyone who was cured did so graciously. Three women who had been praying at the site for a year without a cure complained that Eugenius, who approached the shrine with constipation, was cured immediately following his incubation near the shrine itself. The martyrs then appeared to the women and told them it was not up to them to decide who was cured and when.[39]

Although this was a particular healing cult, from the fourth and fifth centuries, records show that many churches served as hospitals, and different saints were called upon for different ailments such as eye, ear and throat diseases, or general sickness.[40] In the tenth century the relics of Cyrus and John were moved to the Church of Cyrus and John south of Qasr al-Sham, in Old Cairo which reduced the appeal of this church as a centre of healing.

### *Abu Mina*
The complex of Abu Mina in Maryût was 45 kilometres south-west of Alexandria and dedicated to the Egyptian Martyr Menas who was buried at

the site. According to legend, when Menas' remains were being transported to Egypt from Asia Minor, the camel refused to travel further than Abu Mina. The church was founded in 363 CE and the relics were buried in an underground chamber until they were relocated in the fourteenth century to the Church of Saint Menas in Cairo. At 14 metres wide it was the biggest church in Egypt when it was built. The church is likely built upon an older cult centre as several statues dedicated to Horus-Harpocrates were discovered at the site. The site was destroyed in 619 CE by Persian invaders, and archaeological evidence suggests it was burnt to the ground. After they left in 629 CE, the main buildings were repaired but it was not long before the Arab invasion in 639 CE which meant the site was handed over to the Coptic Church, changing the focus of the site with the martyr church being rebuilt in 755–768 CE.

Throughout the fifth and sixth centuries the complex grew to better cater for the pilgrims including accommodation for staff. The churches at the site were not Egyptian in style and this was thought to be in order to cater for international pilgrims. Much of the accommodation, in the form of small rooms, was near the burial of Menas as this would be most beneficial to the pilgrims staying at the site hoping for incubation treatment. There was one semi-circular building here, which accommodated the sick for longer periods of time, and each room was the same distance to the shrine. Other, cheaper accommodation meant the pilgrims lay under a portico rather than being allocated rooms. This area was like a town at the time with streets, arches, squares and public buildings.[41] The main street leading to the spiritual centre, ran north to south and was used by all the pilgrims. The street itself narrowed considerably as it reached the centre, essentially pushing the pilgrims almost into single file. The centre comprised a pilgrims' court, which was colonnaded with a row of shops.

The main focus of worship was the church directly above the burial of Menas, but at the end of the fifth century the newly built Great Basilica was another important aspect of the site. Another addition to the first church was a piscine, a hall where people were baptised. Two or three pilgrims could enter the baptismal font at any one time, and it was clearly an important aspect of the visit to the site as another font was later added to increase capacity. Prior to baptism the pilgrims were required to bathe in the Roman-style baths, which were also used by pilgrims on Maundy Thursday to purify before Easter. There were two baths, and it is thought they may have been used to separate the sexes.[42]

Many pilgrims who were cured of their illnesses or had their problems solved often left a gift for the church. One pilgrim John from Hermonthis (770/80 BCE) who came here looking for a cure for the fatal illness his child was suffering from was so relieved when the child was cured that he gave the child to the church in thanks. It is surprising the church accepted the child. Most pilgrims left money, icons, lamps, curtains, flowers, wreaths or livestock to be slaughtered for the poor, depending on their wealth.[43]

Participating in a pilgrimage was a spiritual experience, but many also hoped to take something with them as a souvenir. For many this may have been in the form of sacred oil which had been presented to a relic or poured over a sacred item or they gouged stone out of the walls removing the dust.

From the site of Abu Mina there were specific souvenirs which could be a small figure of a horseman holding shields, or a pregnant woman which was treated as an amulet for pregnant women or those who were infertile. However, the most common souvenir was a small Menas flask with oil from lamps within the shrines which was known as martyr oil. Studies of the remnants at the bottom of flasks show it contained a lot of suspended incense. These flasks were decorated with an image of Menas between two camels with their heads bowed towards his feet on one side to record how his relics arrived at the site, and an inscription or sometimes an image of Thecla on the other side who was thought to be specifically for female pilgrims and was a way of gendering the experience.

Thecla was a disciple of Paul and appears in the *Acts of Paul and Thecla* and was a popular figure in the early church. In Antioch where she had followed Paul, she was nearly raped by a man called Alexander. She managed to escape and humiliated Alexander. He held some authority with the governor of Antioch, and got her thrown into the arena with lions, bears and bulls. She survived this too, her second martyr ordeal, and many of the images on the Menas flasks show her with an animal in reference to it.[44]

The connection between Theclas and Menas is not clear, and it is thought there may have been a shrine dedicated to Theclas close to the Church of Saint Menas. This is supported by one of the legends about him which has him rescuing a pilgrim to the shrine of Thecla from being sexually assaulted outside.[45]

*Nitria, Wadi Natrun*
Throughout the fourth century many aspiring hermits visited the monastery of Nitria in the Wadi Natrun. It had been founded in 330 CE by Ammon as a place for solitary monks to reside. However, their reputation soon attracted thousands of visitors changing the nature of the monastery to accommodate the tourists. The monks became less solitary and built a community with regular church services for the monks and visitors as well as merchants and bakers on site. It lost popularity through the fifth and sixth centuries and was completely abandoned by the middle of the seventh century.[46]

## Shrine of Paese and Thekla in Busiris

Another popular shrine was that of the martyrs Paese and his sister Thekla, in Busiris. They had been imprisoned for being Christians and in the text regarding one of their visions the whole basis of the shrine as a place of healing is presented.

> The Saviour said to them 'Fear not; it is I who will protect your bodies and I will cause a martyr shrine to be built for you in my name; and whoso shall give an offering to your shrine, I will fill his house with every good thing on earth; and I will cause my angels to protect their bodies and their souls in the aeons of the light ... And I will set my blessing and my peace in the place where your bodies shall be laid. And behold, I have set the angel Raphael to minister to your shrine; and great numbers of sick people suffering from diverse diseases shall come to your shrine, and obtain healing, and go home in peace.[47]

As discussed above, this idea of a shrine being built to house relics which have power was common in the early Christian period, and this site was a popular one.

Some sites in Egypt which did not have obvious Biblical connections were still a draw, and the pilgrims created a narrative which fit into the Christian doctrine. For example, the pyramids of Giza, which do not fit into the Holy Family's Flight, and no obvious Biblical connections, from the medieval period through to the fifteenth century were believed to be the Granaries of Joseph. These are famously depicted in Saint Mark's Basilica in Venice (FIG 11) which was built in 1094 CE, although the five pyramids depicted complete with windows do not resemble those at Giza or indeed any pyramids from Egypt. However, by the time the basilica was built this idea was well-established. When Bernard the Wise (c. 840–900) travelled to Egypt in about 870 he stopped off at the pyramids of Giza firmly believing them to be the Granaries of Joseph.

> they were made by the pharaoh in the time when Joseph, the son of Jacob, was the governor of the kingdom of Egypt. They were made for containing and preserving the grain for the dry period which Joseph had prophesized would come to the kingdom of Egypt, according to the dream of that king pharaoh, as is written more fully in the text of sacred scripture.

Pero Tafur, from Spain (1436) added that the granary was filled by beasts of burden which walked up the side of the pyramid and emptied their grain loads through a window at the top.[48] Philp Melanchthon believed they were built by Israelites when they were enslaved in Egypt. A Dominican monk from Zurich, Felix Fabri (c. 1441–1502) who travelled to Egypt in 1483 with a group of German barons on their way to the Holy Land thought the idea of the pyramids being the Granaries of Joseph to be ludicrous as he denounced, 'the falsity of the opinion of the vulgar who declare that these were the Granaries of Joseph' and instead identify them, quite rightly, as tombs. He was way off, however, when observing the Sphinx, whom he believed to be female and dedicated to Isis.[49] As part of the tour of Egypt, Fabri also visited Alexandria and commented on two obelisks which were to become known internationally as Cleopatra's needles.

The pyramids did appear to be on the route of travellers to Egypt, as well as other ancient sites in between the spiritual visits. For example Ibn Jubayr, an Andalusian traveller (1145–1217) visited Egypt in 1183 and as well as visiting the pyramids at Giza and Alexandria also visited the village of Askun, south of Cairo on the east bank of the Nile, where Moses the interlocutor was born, and apparently where his mother set him on the Nile in a basket.

> On the day of our sailing, and that following, we observed to the west of the Nile and on our right. The ancient city of Joseph the Truthful – God bless and preserve him – where is the prison in which he was confined (Gen 39:20) which is now being demolished and its stones removed to the citadel being built at Cairo.[50]

Burchard of Strasbourg visited the pyramids before heading off to the tree of Matariya, in 1175 and records the experience:

> two monuments of square plan, built of very large stones of marble and other materials; they are admirable constructions, an arrow's distance from each other ... and of the same height and width. Their height is that of a very strong shot of an arrow; their width, of two such shots. A mile from Cairo (Matariya), there was a garden of balm trees, irrigated by a sacred fountain, at which the Virgin Mary was said to have washed the infant Jesus.[51]

Christian pilgrimages were just a small number of journeys made throughout Egypt in the medieval period and were very specific in nature. There were, however, other travellers like Ibn Jubayr who visited a number of sites throughout Egypt and recorded their experiences at this time, with diverse reasons for travelling including pilgrimage, search for knowledge and treasure hunting.

Chapter 5

# Medieval Tourism

Egypt was a Christian country between 391 CE, when emperor Theodosius (347–395 CE) closed all the pagan temples, and the Arab conquest in 640 CE, when General Amr Ibn-al-As (d. 663) took Alexandria. From that moment Egypt was, 'isolated from the Western world',[1] as the new rulers closed off contact with the Byzantium East and the Christian West[2] and the population converted to Islam. Penalties were issued to those who did not convert and by 1000 CE more than half the population were Muslim, with the majority of the remainder converting shortly afterwards.

The Coptic language was another casualty of the Arab conquest, as Arabic became the national language, and it was limited to the few churches that remained in use. The Coptic language was the last link to ancient Egyptian society and language. By the twelfth century, only a few rural communities still spoke Coptic and very few words were absorbed into the Egyptian Arabic dialect. However, although today Islam is the dominant religion with between 85–95 per cent of the population being Sunni Muslims, there are still between 5–15 per cent Coptic Christians with many of them still understanding Coptic as it is spoken in church services.

Another major change following the conquest was moving the capital city from Alexandria, where it had been since Alexander the Great, to Fustat, just south of Memphis which was to be the foundation of modern Cairo. Alexandria was relegated to a minor port and many of the great monuments fell into disuse and ultimately piles of rubble.

## Travellers

This period of history is seen as one clouded in mystery, which has led to such Eurocentric views as Ulrich Haarmann in 2001 stating: 'Any continuity from ancient to Islamic Egypt was irretrievably and doubly cut off, first by the adoption of Christianity in Egypt in the fourth century and then, three centuries later, by the Islamic conquest.'[3] This attitude could be due to the difficulty in accessing resources as much written evidence is still in manuscript form, in varying global collections, and is mostly unpublished, at least in non-Arabic languages. This balance is slowly being addressed in the West, with more work being published on what Okasha El Daly refers to as the *Missing Millennium*.

These sources show that the medieval period was just as active with travellers as previous eras, and many Europeans were very much aware of Egypt and its monuments from their study of a combination of classical texts and Arabic works which were being made available. For example, Constantine the African, an eleventh-century physician, translated a number of Arabic medical texts and is said to have travelled to India, Ethiopia and Egypt.[4]

However, neither Muslims nor European Christians understood the ancient Egyptian culture and viewed it as idolatrous. Historian, Charles Burnett comments:

> although much accurate information in classical sources was available to medieval scholars, it was used selectively, and there was a bias towards regarding the Egyptians as idolatrous, as the foil to the ancient Hebrews, and at best as holding quaint ideas about the gods, which should rather be interpreted as entertaining myths.[5]

Following the Arab conquest there were few opportunities for Western tourists to travel to Egypt and even European merchants travelling to Cairo were told it was too dangerous for them to venture any further south. It was not the same, however, for Arabic merchants who were free (and relatively safe) to travel throughout the country, and, indeed, the Quran actively encouraged Muslims to study ancient cultures.

The native Egyptians, likewise, were, as a whole, uninterested in the monuments for very little other than as quarries and recycling materials from temples, tombs and pyramids became the norm. The white Tura limestone cladding on the pyramids at Giza had already been removed to some extent by the Romans who ground it down to make mortar, but during the early Arab period the rest was stripped and used to build the city of Cairo. Only the tip of Khafra's pyramid retains any of the original stone cladding giving some indication of how it may have looked. In the fifteenth century, the capstones (pyramidion) of the other Giza pyramids were lying on the ground at the base of the pyramids and Egyptian scholar Jalāl al-Dīn al-Suyūṭī (1445–1505) commented that the wind must have blown them down. At the start of the twenty-first century, archaeological excavation uncovered one of these capstones near the pyramid of Khufu.[6]

## Treasure Hunting

With the break in the oral history between the ancient Egyptians and the Muslim invaders, as well as a lack of interest in the written histories left by Greek and Roman travellers, the true history of the monuments of Egypt was gradually lost. However, the human mind will always try to understand what they are looking at, and therefore during the medieval period marvellous ideas abounded.

For example, the pyramids were believed to be full of gold which attracted treasure hunters (matalibben) to Egypt. Jalāl al-Dīn al-Suyūṭī in his treatise on the pyramids wrote that:

> In the western pyramid he made thirty treasure chambers, and filled them with abundant wealth, (various) instruments, and images made of exquisite jewels, as well as fire iron tools, rustproof weapons, glass (of such excellent quality) that (it) would bend and yet not break, strange talismans (various kinds of simple and compound drugs, deadly poisons, and other things.[7]

Treasure hunters were, as a rule, not sole travellers but worked as a team and targeted one particular area at a time.

It became such a draw that ruler, Ahmad Ibn Tulun (868–905) started to monitor the activity and made it necessary to obtain treasure hunting licenses whilst taxing the spoils.[8] Nasir-e Khisraw (1004–1070) records:

> Matalibi is what they call the people who dig for buried treasure in the graves of Egypt. From the Maghreb (Morocco) and the lands of Egypt and Syria come people who endure many hardships and spend a lot of (their own) money in those graves and rock piles. Many a time buried treasure is discovered, although often much outlay is made without anything being found. Whenever anyone does find something, one fifth is given to the Sultan and the rest belongs to the finder.[9]

Ancient Egypt was believed to have been a wealthy nation, and the attraction of hidden treasure was an exciting prospect. In the story of *Pharaoh and Moses* in the Quran, the wealth of pharaoh was emphasised as was that of Qarun, a member of the tribe of Moses. This treasure was said to have been buried in Egypt and it was thought that during a low Nile, the throne of the pharaoh reappeared at the Island of Gold in Giza (Jazirat al-dhahab).[10] Considering, in the same way Christians believed the Bible at the time (see chapter 4),

Muslims believed the Quran to be the truth and therefore did not question the existence of buried gold.

Many treasure hunting guides were written, including *The Ultimate Desires on Precious and Hidden Treasures* MS Arabe 2764) which led the treasure hunters to Wadi Digla (Ma'adi, Cairo), as well as to the Valley of the Kings, and folio 49a discusses the Cave of Cats, where there were thousands of cats laid out on shelves in a catacomb and is thought to be Maabda, near Asyut.[11] Another guidebook, the fifteenth century *Book of Treasured Pearls and Hidden Secrets on the Indications, Cachets, Burials and Treasures* told the treasure seeker where to look for treasure and the necessary spells to overcome the spiritual guardians of the tombs and temples.[12]

The authors of such books (and, indeed, the treasure hunters themselves) believed in these spiritual guardians and took the spells seriously with one prescribing twenty-one days of meditation in isolation. Once this period was over, there would be an appearance of:

> a tall dark servant with large head, riding a horse and [with] a huge lion. He will speak to you but do not answer him. After 35 days, a person will appear, with a dog's face and a human body. He will greet you, do not answer him and he will go away. On day 42, seventy men wearing green shall greet you and you shall answer their greeting. They will say that whatever you demand, they have. ... On day 47, a white city wall will appear to you. It has a great army of cavaliers occupying the valley and the mountain and their noise reaches the horizon. Then, tents will be set up at the gate of this city.[13]

Within the first of these tents the traveller would meet the Imam and several spirits, to whom he would request that they let him pass and enter all the places he wanted to. The imam would do so whenever the traveller repeated the spell.

The Arab visitors were a superstitious bunch, and Al-Idrisi talks about a group of treasure hunters who entered a pyramid in the twelfth century. One member of the group got lost and after three days they thought they would never find him. However, his head suddenly materialised through a wall and his skin was red as he shouted, 'this is the fate of those who violate the sanctity of kings in their homes'. He had initially been speaking in an unknown tongue which they got a local monk from a monastery to interpret indicating that Coptic Christians were still believed to understand the ancient Egyptian languages.[14]

The guidebooks were particularly specific about where to find treasure and one is left wondering why the authors did not go and discover it themselves. One writer of treasure hunting guides stated he wrote them, not for personal gain but to raise money for charitable causes.[15] They include detailed maps and plans of sites and tombs, which have since proved useful to archaeologists.[16] For example, Al Idrisi records a book he received which describes seventy pyramids on the Muqatam Mountain:

> Walk east until you pass by the area with lots of black roots like wood [*petrified forest*] till you find a cave … and until you get to a high mountain leading to the tombs. Look down the valley and you will see a mound stretched in the valley of a mountain, and nearby seventy pyramids of black stones. Measure from the front of each pyramid seventy-one feet [*21.6 m*] and dig. Go down seventy steps cut out in the mountain, you will find closed houses to right and left. Open carefully; you will find money, gems and inlaid jewellery.[17]

This petrified forest is in an area east of Ma'adi, Cairo called Qatamiya which was made into a nature reserve in the early 2000s when officials realised much of the petrified wood was being removed for building works in Cairo.

In 1900 a conservator at the Cairo Museum complained that this treasure hunting book was still being consulted and, 'this work had been responsible for destroying more ancient monuments than either war or the onslaught of the centuries'. Gaston Maspero oversaw the publication into Arabic of a guidebook not long after hoping wide availability would discredit it somehow, rather than providing a wider audience.

Throughout the medieval period, treasure hunting was a legitimate reason to visit Egypt. Abd el-Latif (1162–1231), a physician from Baghdad, recorded such a trip:

> A credible person told me that, joining once in a search for treasures, near the pyramid, his party found a tightly sealed jar, on opening which and finding honey, they ate it. One of them noticed a hair that had stuck to his finger: he pulled it towards him and a small infant appeared, the whole of the limbs of which appeared to have preserved their original freshness.[18]

Other treasure hunting trips were less disgusting and more successful. Jabir Ibn Hayan recorded in his book *Al-Naqd* that one pyramid had thirty pharaonic jars filled with a red elixir, and another pyramid contained precious gems which were so old it was impossible to see what they were.

Eighth-century Caliph el-Mamun, son of Caliph Harun el-Rashid, believed the Great Pyramid to be full of treasure and was determined to enter and claim it for himself. He started by covering the north face of the pyramid with hot vinegar hoping it would be enough to crack the rock. But when that did not work, he resorted to use a battering ram to force an entrance – this is the same entrance which is still used by tourists today. It was recorded that when the pyramid was opened, a green jar full of gold was found, to the exact amount used in breaking into the pyramid. Tunnelling through the pyramid the caliph found the queen's chamber and the king's chamber. The former was full of bats and a number of Late Period mummies:

> He [al-Qaisi] was informed that those, who went up there in the time of Al Mamoon, came to a small passage, containing the image of a man in green stone, which was taken out for examination before the caliph; and that when it was opened a human body was discovered in golden armour, decorated with precious stones, in his hand was a sword of inestimable value, and above his head a ruby the size of an egg, which shone like fire, and of which Al Mamoon took possession.[19]

This body itself sounds more medieval than ancient Egyptian, and indeed rubies were not known in Egypt until the Roman period so this body was definitely not an Egyptian king, and one could even suggest el-Mamun made it up. al-Idrisi himself claims there was little else in the pyramid other than a few decayed remains.

However, there had been tales about the pyramids being full of treasure since Al-Mu'udi was writing in the tenth century. He claims to have found an ancient book which outlined the wonderful treasures to be found in the structures. The ruler Al-Ikhshid gave permission to excavate and uncovered wooden coffins, painted in elaborate designs and covered in bitumen. Their eyes were inlaid with semi-precious stones, and some had gold and silver masks. There were also jars of remaining paint nearby which were ground down and burned as a form of medicine. As plausible as this sounds the same account was then recounted by Al-Bakri (eleventh century) and then by Al-Idrisi (twelfth century) meaning anything that had been there originally would have been depleted over the two preceding centuries.

Not all treasure hunters, however, were looking for gold, precious metals and jewels but were also looking for medicinal items, such as kohl which was said to cure blindness, or they were hunting for the ancient Egyptian books of wisdom,[20] even though they would have been unable to read them had they been discovered. As with the record of the distinctly medieval sounding body, not all records of the books found were entirely truthful and many alchemists claimed to have discovered books written by Hermes himself, in secret underground tunnels of Egyptian temples.

As we know, however, some of the monuments of Egypt did hold ancient treasure such as the tomb of Tutankhamun, and the depositories at the temples of Tod and Dendera. A goldsmith's workshop had been situated at Dendera, so when Al-Dimishqi (d. 1328) reports that a hoard of precious metals including gold was found by someone digging, it is not outside the realm of possibility.[21] These treasure hunters were simply looking in the wrong place, by focusing on the larger, more obvious monuments like the pyramids and for many it was a hobby which was to ruin them. Ibn Qadi Shuhba, said of the death of Sheikh Muhammad Ibn Mubarak AL-Athari in 1403: 'He was obsessed with treasure hunting, spending all his earning on the search, but never gained any.'[22]

Treasure hunting had become such an important aspect of the local economy that Ahmad Ibn Tulun (868–905) was said to have built his mosque (FIG 12), a state hospital and other building projects using the proceeds of discovered pharaonic treasure. Possibly as much as 4,000 kg of gold.[23] He therefore saw the value in treasure hunting and looking for ancient gold that he created an industry by issuing permits which were overseen by a government department.[24]

His mosque is still one of the oldest mosques in Africa and has been a draw for tourists since its inauguration in 879 CE. Jubayr (1183) described it as, 'one of the old congregational mosques, of elegant architecture, and of large proportions', which had been built as a retreat for people from the Maghrib, (western Barbary and Spain).[25] The layout is formed of a large courtyard of 92 square metres, with an ablution fountain in the centre, and an arched colonnade surrounding the courtyard. The minaret of the mosque which has a square base, topped by a curved shaft is thought to have been modelled on the Lighthouse of Alexandria.

Not everyone agreed with the treasure hunting as a great way to raise state cash. The jurist, Ibn al-Haj (d. 1337) believed treasure hunting to be an illness, and that it went against the teachings of Islam. He thought that many of the books were written purely to make money from the poor and desperate. At the time he was writing it had become so mainstream that no longer was it carried out in secret, but in broad daylight. He also stated that it was so prevalent that if you wanted to have someone's house destroyed you could produce a paper, made to look ancient, claiming there was treasure under the house. The building could then be pulled down.[26]

What did the treasure hunters do with everything they found? Well, of course the precious metals could be used as currency as seen with Ibn Tulun, but there is evidence that 'Umar Ibn 'Abd Al-'Aziz the caliph of Damascus between 717 and 720 collected ancient Egyptian artefacts and showed them to his guests with pride. He had several stone statues which he believed were offerings which had been turned into stone by the God of the Pharaoh Moses as punishment.[27]

This interpretation does seem bizarre with hindsight, but at the time they had no real knowledge of the civilisation or what the objects represented and therefore had to find interpretations which fit into their lifestyle and beliefs. Alchemists for example believed many of the ancient artefacts to be related to lost alchemical knowledge.[28] Astronomers discussing a stela depicting a traditional smiting scene believed it to be a representation of the planets whereas Ibn Umail (900–960), added to the discussion to say it clearly had alchemical meaning.[29]

*Alexandria*
Once Alexandria had been abandoned as the capital city it became a shadow of its former glory and when Laurence Aldersley (1581–86), an English adventurer travelled there he recorded that, 'outside the city walls [was] Pompey his pillar, which is a mighty thing of grey marble, and all of one stone',[30] indicating this monument, then as now dominated the landscape.

Ibn Jubayr, a geographer from al-Andalus recorded his pilgrimage to Mecca between 1183 and 1185 which also included a stop in Alexandria:

> We observed many marble columns and slabs of height, amplitude, and splendour such as cannot be imagined. You will find in some of its avenues columns that climb up to and choke the skies, and whose purpose and the reason for whose erection none can tell. It was related to us that in ancient times they supported a building reserved for philosophers and the chief men of the day. God knows best, but they seem to be for the purpose of astronomical observations.
>
> One of the greatest wonders that we saw in this city was the lighthouse which Great and Glorious God had erected by the hands of those who were forced to such labour as 'a sign to those who take warning from examining the fate of others' (Koran XV, 75) and as a guide to voyagers, for without it they could not find the true course to Alexandria. ... Description of it falls short, the eyes fail to comprehend it, and words are inadequate, so vast is the spectacle.[31]

As well as taking measurements of the lighthouse, Jubayr also went inside, and his description indicates there was a great deal of it left at the time: 'Its interior is an awe-inspiring sight in its amplitude, with stairways and entrances and numerous apartments, so that he who penetrates and wanders through its passages may be lost.'³²

The Lighthouse of Alexandria was clearly a draw for visitors during this period and Abu Hamid Al-Gharnati (1110 and 1117), an Andalusian traveller described it:

> The first tier is a square built on a platform. The second is octagonal and the third is round. All are built in hewn stone. On the top was a mirror of Chinese iron of seven cubits wide [3.64 metres] used to watch the movement of ships on the other side of the Mediterranean. If the ships were those of enemies, then watchmen in the Lighthouse waited until they came close to Alexandria, and when the sun started to set they moved the mirror to face the sun and directed it onto the enemy ships to burn them in the sea. In the lower part of the Lighthouse is a gate about 20 cubits (10.6 metres) above the ground level; one climbs to it through an archway ramp of hewn stone.³³

Jubayr indicates that there may have been a tourist industry alive in Alexandria in the twelfth century as he talks about information which 'was related to' them as well as describing:

> colleges and hostels erected there for students and pious men from other lands. ... The care of the sultan for these strangers from afar extends to the assigning of baths in which they may cleanse themselves when they need, to the setting up of a hospital for the treatment of those of them who are sick, and to the appointment of doctors to attend to them.³⁴

Jubayr continued his journey from Alexandria all the way to Upper Egypt visiting many villages and monuments along the way including Cairo, Akhmim and Dendera. He records crossing the Nile whilst in the Delta in a ferry boat suggesting these were available to hire for those who needed to cross.

Unlike Western tourists to Egypt, Jubayr stopped at several Muslim sites of note and commented that while in Birmah, in the western Delta near Tanta, they 'shared in the prayers in a place called Tandatah, a large and populous village, where we observed a vast concourse being addressed by the preacher in an eloquent and comprehensive discourse'.³⁵

He also visited a shrine in Cairo which had relics in the form of the head of Husayn, the son of 'Ali ibn Abi Talib, son-in-law of the prophet Mohammed:

> It is covered with various kinds of brocades and surrounded by white candles that are large columns; smaller ones are placed, for the most part, in candlesticks of pure silver and of gilt. Silver lamps are hung from it and its whole upper part is encircled with golden spheres like apples, skilfully executed to resemble a garden and holding our eyes in spell by its beauty.[36]

Relics had been popular throughout the Christian period of history (see chapter 4) and was a way of connecting a local place of worship with the Bible, or in this instance with the Quran. Whilst there Jubayr witnessed pilgrims kissing the tomb and throwing themselves prostrate upon it as well as circling it in reverence, crying out and even weeping with ecstasy which he considered an 'awe-inspiring sight'.

Travelling to Cairo in 1326 Ibn Battuta, a Moroccan explorer, talks about the festivals and religious practices he witnessed which were different to those Jubayr witnessed at the tomb:

> On the bank of the Nile opposite Cairo is the place known as al-Rawda (the Garden), which is a pleasure park and promenade, containing many beautiful gardens. The people of Cairo are fond of pleasure and amusement. I once witnessed a fête there which was held for al-Malik al-Nâsir's recovery from a fracture which he had suffered in his hand. All the merchants decorated their bazaars and had rich stuffs, ornaments, and silken fabrics hung up in their shops for several days.[37]

### Akhmim

Even though it is often presented that there was little interest in the pre-Islamic history, many Arab visitors to Egypt visited some of the 'key' sites, the most common throughout the early medieval period being the pyramids at Giza, but also the temple of Akhmim where the remains of a large eighteenth dynasty temple stood. This temple was built in by Thutmosis III (1504–1450 BCE) and had been added to by later kings. Another temple at the site was built during the Roman period. There is little left of this once important site, with the temples being dismantled for building material by the local villagers and the ancient town itself abandoned.

Ancient temples were not automatically recognised as ancient places of worship, perhaps because they differed so much from the Muslim mosques, and Ibn Al-Nadim (d. 920) presented a theory:

> In Egypt there are buildings called barabi made of immensely large great stones. The birba are temples of different designs, and have places for grinding, pounding, dissolving, assembling and distilling, showing that they are built for the craft of alchemy. In these buildings are reliefs and inscriptions in Chaldean and Coptic; their meanings are not known ... the known barabi are the temples of wisdom.[38]

Al-Aqzwini (d. 1283) describes the birba at Akhmim: 'The birba of Akhmim is a temple which has images depicted in the stones, high reliefs, still visible until now.'[39] Ibn Jubayr (1145–1217) visited in May 1183 and described the temple in more detail:

> the most remarkable of the temples of the world ... This great temple is supported by forty columns, beside its walls, the circumference of each column being fifty spans and the distance between them thirty spans. Their capitals are of great size and perfection, but in an unwonted fashion and angulated in ornate style as if done by turners. The whole is embellished with many colours. ... The ceiling of this temple is wholly formed of slabs of stone so wonderfully joined as to seem to be one single piece; and over it all are disposed rare painting and uncommon colours, that the beholder conceives the roof to be of carved wood. Each slab has a different painting. Some are adorned with comely pictures of birds with outstretched wings making the beholder believe they are about to fly away; others are embellished with human images, very beautiful to look upon and of elegant form ... each image has a distinctive shape, for example holding a statuette or a weapon, or a bird, or a chalice, or making signs to another with the hand.[40] There is hardly the space of an awl or needle hole which did not have an image or engraving or some script which is not understood.[41]

This record not only shows Ibn Jubayr's awe at the beauty and size of the temple but is also a valuable record of what the temple once looked like, as progressively throughout the medieval period this temple was quarried and plundered until there was virtually nothing left. Al-Tujibi (d. 1329) noted that since the time of Ibn Jubayr, two of the columns had disappeared, meaning he only counted thirty-eight. He also mentioned in his records that local people visited the temple as they believed the images on the walls were able to perform magical acts if the questioner left offerings, although he believed this to be the thoughts of 'ignorant people'. However, these traditions seemed to be widespread throughout Egypt at the time, and another record from the thirteenth century tells how pilgrims to Biyahmu in the Fayoum left offerings and burnt incense at the feet of the Middle Kingdom statues, as well as drinking the water from beneath their feet for medicinal purposes.[42]

Construction work at the dilapidated site in the 1980s uncovered a Ptolemaic gateway and some statues dated to Ramses II, and the Roman period showing the splendour of the temple once there. They have been displayed in an open-air museum at the site. Further excavations have uncovered the remains of a large temple built by Ramses II.

*Pyramids at Giza*

The pyramids at Giza have been a constant draw for visitors since they were built, but during the medieval period the history had long been lost, meaning there were many theories of why and when they were built. Jalāl al-Dīn al-Suyūṭī (1445–1505) believed they were built before the 'deluge' and that:

> The Pyramids baffle the minds of men of intelligence,
> And the most extravagant dreams become insignificant compared with
>     their magnitude.
> Smother, of triangular build, and lofty,
> Arrows shot to their utmost distance fall short of them.
> I know not – since meditation stops short before them,
> And conjecture is perplexed by their marvel —
> Whether they are graves of heathen kings,
> Talismans against the (ravages of the) sand, or monuments.[43]

Many at the time believed they were built with the aid of magic, showing that such fringe theories are nothing new. Abdullah Muhammad Ibn Batutta (1304–1368/69) believed that Hermes Trismegistus used the pyramid to store all his wisdom to protect it from the Great Flood.[44] This attribution to Hermes had been passed on via the Copts and the Gnostics and played an important part of the (incorrect) understanding of the ancient monuments at the time. Hermes was seen as a figure from the ancient past who was the source of all knowledge and was responsible for building cities, temples, pyramids and writing. The astrologer Abu Mashar al-Falaki (787–886) claimed Hermes:

> wrote for his contemporaries many books ... about the knowledge of terrestrial and celestial subjects ... It was because of his fear that wisdom might be lost that he built the temples, namely, the mountain known as al-Barba, the temple of Akhmim (Panopolis), engraved on their walls drawings of all techniques and their technicians, made pictures of all the

working-tools of craftsmen, and by inscriptions indicated the essence of the sciences for the benefit of those who were to come after him. In doing so, he was guided by the desire of preserving science for later generations and by fear that its trace might disappear from the world.[45]

Saad Allah Abu al-Makarim (twelfth century) wrote that the two largest pyramids were the tombs of Hermes and Agathodaimon and that were the result of hard work and not magic.

> When a man of sense beholds these ruins he finds himself able to excuse in the vulgar their belief with regard to the ancients that their lives were longer than ours and their bodies stronger, or that they possessed a magic rod with which when they struck the stones they leapt towards them. For the modern mind feels itself unable to estimate how much was required in these works of knowledge of geometry, and concentration of thought, and ardour of study, and patience in labour, and power over tools and application to work.[46]

This idea of the pyramids as tombs of Hermes and Agathodaimon had been suggested by A'bd al-Latif al-Baghdadi (1162–1231) who also tried to tie the pyramids with the time scale of the Quran, even though they are not mentioned in the book at all. He participated in the debate about whether they were pre-flood, and he even entertained the idea that the pyramids were even pre-Adam which suggested a pre-Quranic history.[47]

Scholars were greatly attracted to the pyramids and were very keen to provide data about the structures in the form of measurements and geology of the stones. Abu Jafar al-Idrisi (d. 1251), for example, wrote the first detailed study of the pyramids called *Light on the Voluminous Bodies to Reveal the Secrets of the Pyramids*. This included all measurements and locations of the pyramids, as well as a study of previous literature, although had he included the classics like Herodotus, he would have been able to get some historical context for them. It was not really until the fifteenth century that Herodotus was once more gaining in popularity and Italian philosopher, Lorenzo Valla (1407–1457) created a translation in 1452 which was published twenty years later.[48]

However before this translation was available Burchard of Strasbourg visited in 1175 and records the experience:

> two monuments of square plan, built of very large stones of marble and other materials; they are admirable constructions, an arrow's distance from each other … and of the same height and width. Their height is

that of a very strong shot of an arrow; their width, of two such shots. A mile from Cairo (Matariya), there was a garden of balm trees, irrigated by a sacred fountain, at which the Virgin Mary was said to have washed the infant Jesus.[49]

When Laurence Aldersley, the English adventurer, visited Egypt (1581–86), he was rather taken with the pyramids at Giza and their size: 'the monuments bee high and in forme 4. Square and every of the squares is as long as a man may shoote a roving arrowe, and as high as a Church, I sawe also the ruines of the Citie of Memphis hard by those Pyramides'.[50]

Perhaps it was part of a rough itinerary at the time to go from Cairo to Memphis as many travellers seemed to visit in this order. Benjamin of Tudela (1130–1173), a Jewish traveller from Navarre in northern Spain, for example, arrived from the south, through Abyssinia which was an unusual way to approach Egypt. He observed on his travels that the Nile flood coincided with the monsoon rains in Abyssinia. Upon travelling north he visited the pyramids, which he believed were built by magic and then went to Memphis and Saqqara as did al-Idrisi who described Memphis as requiring: 'a half-day's march in any direction to cross the visible ruins. It is a gathering of marvels to confound the mind. The more one looks upon them the greater grows the admiration they inspire; and each fresh glance of the eye upon these remains is a renewal of delight.'[51]

It would seem there may have been a certain element of scholarly credibility in visiting the pyramids when in Egypt. However, not all travellers who wrote about them seemed to have made the trip. Take Abdullah Muhammad Ibn Batutta (1304–1368/69) for example whose records of the pyramids makes it abundantly clear he had not seen them: 'The Pyramid is an edifice of solid hewn stone, of immense height and circular plan, broad at the base and narrow at the top, like the figure of a cone.'[52] And considering that most of the people reading his work probably had not seen them either, he probably got away with it.

It is recorded that during the Fatimid dynasty, the celebration 'Night of Fire' was kick-started by lighting a fire on top of the pyramid here. Jalāl al-Dīn al-Suyūṭī and the Scribe of Salahdin, El Emad al Isfahani (d. 1201) both record the Sabians at the site. They were a Turkish religious sect mentioned in both the Quran and Bible. Isfahani (d. 1201) recalls they were 'dressed in the manner of Iraqis and Syrians wearing head scarfs'. The Sabians visited the pyramids as a pilgrimage, believing them to be the tombs of Seth, Hermes and Sab, son of Hermes who they honoured by burning incense and sacrificing black calves and white roosters.[53] It is thought this belief could be a corruption of the text of Pliny which records that King Harmakhes was buried here, which was a corruption of the name Hore-em-khat, the name of the Sphinx.[54]

*Sphinx*
The Sphinx, like the pyramids was a prominent feature on the Giza plateau, and the Arab travellers referred to it as 'Terrifying', 'Great Terror', or 'Father of Terror' with Jubayr (1183) claiming it to be 'a strange figure of a man of fearsome aspect'.[55] Other writers however, like Al-Baghdadi comment on its benign smiling face. al-Idrisi (1100–1165) described it as, 'very beautiful ... and its mouth carries the stamp of grace and beauty. It could be said that it smiles graciously'.[56] At the time, the face was a vibrant red and was still intact, although Jalāl al-Dīn al-Suyūṭī explains how 'nothing is seen above the surface of the earth but the idol's head and neck', demonstrating that at this time it was buried. He goes on to say that the Sphinx was a talisman to protect the cultivated area from a build-up of sand.

Christian visitors, on the other hand, viewed the Sphinx as an 'immense idol of stone which had the shape of a woman'.[57]

*Heliopolis*
Another popular place to visit in the medieval period was the Grand Shrine of Heliopolis which was dedicated to seven deities connected with the seven heavenly bodies or planets with the sun god at the head. It was a pilgrimage site for Egyptians and pilgrims from around the world, where they participated in daily prayers which took place at dawn, noon and sunset. The site was known as Ain Shams in reference to its connection with the sun god.[58] Al-Idrisi (1100–1165 CE) states this site was popular with Sabians as well as local Egyptians which is corroborated by Al-Maqrizi (1364–1442 CE) who emphasises that centuries later both Egyptian and international pilgrims came here.

It remains one of the oldest suburbs of Cairo although there is very little there to show its ancient past, or what may have been extant during the medieval period. All that remains is an obelisk of Senusret I (1971–1926 BCE) which was no doubt visible to the medieval visitors.

*Abu Sir*
Abu Sir just southeast of Cairo, was an important pilgrimage site for many Muslims. The sijn yousef or Prison of Joseph, where it was believed Joseph (an Egyptian prime minister) received a number of divine revelations when he was incarcerated here was located at Abu Sir.

This was also the site of the ancient cult of Imhotep, and it was thought that this god answered pilgrims' requests at the site. The general locale was also associated with the oracle of Hermes Trismegistus, a popular writer with Muslim scholars. Therefore it seems like the Muslim visitors were continuing the traditional oracle practices at the site which had been prevalent since the pharaonic period (chapter 7).

Medieval Muslims visited on pilgrimage and there were annual festivities which lasted for three days.[59] Ibn Umail, visiting in the tenth century CE was an alchemist and records visiting a chapel on more than one occasion with his friends. One Arab author (fol. 58b) describes that whilst in the Abu Sir necropolis they saw the Manahat Al-Qitat (cat burials) which was an interesting anomaly for the Muslim visitor.[60]

*Dendera*

When continuing his travels down to Dendera, Ibn Jubayr (1145–1217) did not visit the temple at Dendera and simply records that 'we were told that it has a great temple' and that it was 'even more magnificent and larger too than' the temple at Akhmim.[61]

The temple was believed to have been designed with astronomical features such as 180 niches through which the sun entered the temple; a different niche every day.[62] The astronomical ceiling at Dendera records the journey of the sun through a year and is divided into two halves of 180 days each which may have guided some of this thinking.

Dendera was thought to be a depositary for treasure and Al-Dimishqi (d. 1328) recorded that a man digging illegally here, found precious metal hoards and when arrested by the authorities he had one hundred sacks of treasure with him.[63]

*Luxor*

Abdullah Muhammad Ibn Batutta (1304–1368/69), a Berber Maghrebi scholar and explorer, also made it as far south as Luxor and visited Luxor Temple although he only talks about Abu-al-Haggag Mosque (FIG 13) which is built behind the first pylon. Whilst Jason Thompson claims he, 'was deliberately blind to antiquities'[64] this is unfair, as he was simply recording what interested him. Many modern tourists to Luxor Temple do exactly the same thing but in reverse, ignoring the mosque, despite its age, in favour of the pharaonic temple which is the reason they have visited the site. The mosque was built in the thirteenth century and was registered as an antiquity in 2007. It was built by Abu Al-Haggag (The Father of Pilgrims) Yousef Ibn Abdel Raheem (1150–1245 CE) who is also buried inside.

One of the guidebooks talks about the temple of Coptos (Qift) which is just north of Luxor, and describes a wonderful painted ceiling showing a full-length figure of Miriam (Mary).⁶⁵ This would have been an image of the sky-goddess Nut and it is interesting to see she had been equated with Mary at this time. The ceiling at this temple is no longer in existence, so it is an insight into what the remains looked like in the fourteenth century and indeed could be the only description left of it.

*Colossi of Memnon*
The Colossi of Memnon have stood on the same sight, calmly looking over the landscape, since they were built by Amenhotep III (1386–1349 BCE). However, in the fourteenth century they were referred to as a king, Shama and his lover Tama. The statues, due to a crack in the stone made moaning sounds at dawn, and in this period they were seen as whispering sweet nothings to each other.⁶⁶

## Destruction

The Arab conquerors have often been blamed for a great deal of damage to the ancient monuments – some accurately and some not quite so much. For example, Caliph Umar and Amr Ibn-el-As who took Alexandria in 642 CE are often held responsible for the destruction of the library of Alexandria. It is recorded that Amr Ibn-el-As approached the caliph and asked what to do with the library. The caliph responded:

> as to the books which you have mentioned, if they contain what is agreeable with the book of God, what is in the book of God is sufficient without them; and if they contain what is contrary to the book of God, there is no need of them: so give orders for their destruction.⁶⁷

Apparently, the books were then burned as fuel to heat the baths and within six months the library was empty, other than the works of Aristotle. However, as discussed in chapter 6 the library was likely to have been destroyed during the reign of Julius Caesar in 48–47 CE.

As had been traditional since the pharaonic period, temples and monuments were quarried for their cut stone and the Arab rulers simply continued this practice. Columns, in particular, were popular and were removed for the building of mosques and churches as well as other building projects. For example, Al-Baghdadi records that more than 400 pillars in Alexandria, like Pompey's Pillar, had been broken up and piled on the beach as a sea defence to protect the inhabitants from the waves but also to prevent enemy ships from landing. This destruction was said to have happened under the leadership of Ayyubid Sultan Salah Al-Din Yousuf Ibn Ayyub (Saladin) between 1169–93 and was condemned by Al-Baghdadi as 'a work of childish folly committed by those who do not distinguish between a beneficial act and a heinous one'. Is this any more ridiculous than the modern Egyptian government dumping concrete blocks as a tide break over the remains of the royal palace in the twentieth century (chapter 1)? He further condemns those who demolished monuments at Saqqara in the hunt of the copper used to clamp stones together.[68] al-Idrisi also comments that under Saladin, a number of small pyramids at Giza were destroyed in order to use the blocks in Cairo's walls and bridges.

The nose of the Sphinx is another item of controversy, where one story states it was Napoleon and his troops who used the Sphinx for target practice and destroyed the nose, whereas other stories state it was Sa'im al'Dahr who destroyed the nose in protest to the local Egyptians making offerings to the Sphinx which he viewed as blasphemous.

As we travel back in time, we will see that every era of Egyptian history is responsible for damage to the monuments, whether intentionally, accidently or in the name of progress.

Chapter 6

# Graeco-Roman Tourism

The arrival of Alexander the Great in Egypt in 332 CE saw the start of the Ptolemaic period, which ended with Cleopatra VII (51–30 BCE), when Egypt was annexed by Rome. This was the start of an interesting time in travel and tourism.

During the Ptolemaic period there was a large influx of Hellenes to Egypt, and large cities like Alexandria became multi-cultural with the Hellenes and the Egyptians living side by side. As can be expected with a large ex-pat community, there was an increase in tourism by the Hellenes to traditional Egyptian sites, which were considered exotic and 'other'. Diodorus Siculus travelling in approximately 60 BCE and Strabo (24 BCE) both comment that Egypt was a fascinating place as the customs were paradoxical to the Greek's, with items of note being the treatment of animals, funerary customs, monuments and the Nile itself. However, Egypt had been part of the Greek literary tradition for at least a century before Alexander arrived, meaning it was an important part of their cultural history and they had already been travelling in smaller numbers prior to this period.

## Guidebooks

The existence of travel guidebooks at the time show there was a taste for seeing the wider world although there was no Greek or Latin word for tourist. However, tourism clearly existed at this time, and, to a certain extent, a rudimentary tourist industry. Guidebooks themselves were locational history books rather than guides, providing a narrative history of culture, customs, architecture and religion in relation to key sites. They were designed to guide you around the country with the goal of increasing knowledge and not falling foul of local custom. Aelius Aristides (117 CE–181 CE) is recorded as saying that the whole function of a guidebook was to help the reader with laws and customs although, 'there is no need anymore ... Since the civilised world is now open and the law is everywhere the same'.[1]

One of the earliest Greek travellers and chroniclers to Egypt was Herodotus who wrote his *Histories* in approximately 460 BCE. He travelled around Egypt (including Memphis, the pyramids, Heliopolis, Hawara, and Elephantine although probably not Thebes) and based his *Histories* on a combination of first-hand experiences as well as relying on priests and tour guides to provide him with the information he required; essentially 'sightseeing and oral tradition', rolled into one.[2] He mentions at times the use of an interpreter and one wonders if this was something offered to other travellers in Egypt if they requested it. Some of the information Herodotus was given was not accurate and it is thought that, in some cases, the priests told him stories which fed the Greek biases he held against the Egyptians and their culture.

Later writers took great pleasure in showing the inaccuracies or 'lies' in his work to raise the reputation of their own. Diodorus Siculus (60 BCE), for example, is rather disparaging when he claims he will 'omit from our history the tales invented by Herodotus and certain other writers in Egyptian affairs, who deliberately prefer fable to fact'.[3] However, Diodorus and others used the same research techniques (even drawing on the text of Herodotus) which as time progressed became less and less accurate as the people who remembered the stories and could read the hieroglyphs slowly disappeared. Unlike in the modern world, there were no other sources upon which to double check facts.

It has been said that *Histories* (Book 2) was essentially the first guidebook on Egypt, highlighting worthy places to visit, interesting cultural things to look out for, as well as explaining aspects of the landscape, climate and travelling times:

> Southward of Heliopolis the country narrows. It is confined on the one side by the range of the Arabian mountains which run north and south and then continue without a break in the direction of the Arabian Gulf. In these mountains are the quarries where the stone was cut for the pyramids at Memphis.[4]

Herodotus himself even refers to many Greeks who travelled to Egypt, showing that at this point in history travelling was not uncommon. This idea of visiting sites for their worthiness alone was one that endured and the term 'worthy' or a 'spectacle' is repeated in many other guidebooks of the era such as Strabo's *Geography* in reference to the pyramids, Philae, the Valley of the Kings, and the oasis. These phrases can also be seen in the graffiti and inscriptions left behind by the tourists, showing they were visiting them for tourism and their 'worthiness' simply as a place to visit.

What is particularly fascinating about some of the classical accounts is that they were often incorrect in their assumptions about what many of the monuments were about and what they may have originally looked like or been

Funerary boat of Khufu, Giza. (*Photograph by the author*)

Dr Raghab's Pharaonic Village, Cairo. (*Photograph by the author*)

Kom el Dikka open air museum, Alexandria. (*Photograph courtesy of BKB Photography*)

Karim Nile Steamer. (*Photograph courtesy of Marc Ryckaert, Wikimedia Commons*)

A room at the Mena House, Cairo, (*Photograph courtesy of Library of Congress, Prints & Photographs Division* [*LC-DIG-matpc-03826*])

Winter Palace, Luxor. (*Photograph courtesy of Ian Pudsey, Wikimedia Commons*)

Entrance to the Great Pyramid of Giza. (*Photograph courtesy of BKB Photography*)

Abu Simbel. (*Photograph courtesy of BKB Photography*)

Kingston Lacy obelisk, UK. (*Photograph courtesy of BKB Photography*)

Hanging Church, Cairo. (*Photograph courtesy of BKB Photography*)

Granaries of Joseph, Venice. (*Photograph courtesy of Rob Hurson, Wikimedia Commons*)

Ibn Tulun Mosque, Cairo. (*Photograph courtesy of BKB Photography*)

Abu-al Haggag Mosque, Luxor Temple. (*Photograph courtesy of Roland Unger, Wikimedia Commons*)

KV9, tomb of Ramses V and VI, Valley of the Kings, Luxor. (*Photograph courtesy of Dmitrii Zhodzishskii, Unsplash*)

The mortuary temple of Sety I, Abydos. (*Photograph courtesy of BKB Photography*)

Colossi of Memnon, Luxor. (*Photograph courtesy of BKB Photography*)

Gouges at Karnak Temple, bottom left of image. (*Photograph courtesy of BKB Photography*)

Sphinx, Giza. (*Photograph courtesy of BKB Photography*)

Travelling barbers in the tomb of Userhet, TT51, Luxor. (*Photograph courtesy of Metropolitan Museum of Art, Wikimedia Commons*)

Overnight at Karnak Temple. (*Photograph courtesy of Ron Porter, Pixabay*)

used for. For example, Strabo describes Abydos as a large city, when in fact we know it to be a complex of temples, and Diodorus described the long-lost wonders of Thebes: 'The richness and finishing of the ornaments corresponded with its grandeur. Several kings contributed to embellish it. It still subsists, but the gold, the silver, the ivory and the precious stones were carried off, when Cambyses set fire to all Egyptian temples.'[5]

Alongside the guidebooks were paradoxographical works – that is the listing of international wonders whether the author had been there or not. One such writer was Pomponius Mela, a geographer during the reign of Caligula (37–41 CE) who in his works *De Chorographia* lists several notable sites in Egypt. There is doubt that he visited Egypt at all. These books read a little like guidebooks, describing the seasons and the wonders that can be seen but are more verbose and it is thought these works were more likely to be used by poets and scholars looking for unusual names and practices rather than tourists trying to plan their itinerary.

In addition to using guidebooks like Herodotus' *Histories*, Strabo indicates there was an industry of tour guides at the sites.

> At Heliopolis we saw large buildings in which the priests lived. For it is said that anciently this was the principal residence of the priests, who studied philosophy and astronomy. But there are no longer either such a body of persons or such pursuits. No one was pointed out to us on the spot, as presiding over these studies, but only persons who performed sacred rites, and who explained to strangers [the peculiarities of] the temples. (17.1.29)

Unfortunately, like today's guides, the information provided was not always accurate and Strabo laments this. He had been told at the pyramids, for example, that the stones on the floor were fossilised lentil husks which were eaten by the ancient pyramid builders. Whilst it makes an interesting story one wonders if Strabo genuinely believed it.

In very few cases do we have the names of these tour guides and one known guide was Chaeremon of Alexandria who took Aelius Gallus down the Nile and filled him in on the history of the sites. However, Strabo tells us he 'pretended to some knowledge of this kind, but he was generally ridiculed for his boasting and ignorance'.

## Alexander the Great

Almost from the moment he arrived in Egypt, Alexander the Great travelled the country and could be cited as the instigator of an increased tourist trade in Egypt. He started his journey by stopping at the small fishing village of Rhakotis, just opposite the island of Pharos. Plutarch (approx. 66 CE) records the foundation of Alexandria on this spot:

> After Alexander had conquered Egypt, he was anxious to found a great and populous Greek city there, to be called after him ... [A]s he lay asleep he dreamed that a grey-haired man of venerable appearance stood by his side and recited lines from the Odyssey.
>
> 'Out of tossing sea where it breaks on the beaches of Egypt
> Rises an isle from the waters: the name that men give it is Pharos.'
>
> Alexander rose the next morning and immediately visited Pharos. ... he declared that Homer, besides his other admirable qualities, was also a very far-seeing architect, and he ordered the plan of the city to be designed so that it would conform to this site.[6]

It was here that he built a great city which was named after himself – Alexandria. The site was chosen as it had access to fresh water from the Canopic branch of the Nile, robust transportation links with the Mediterranean, and was protected by Lake Mariut to the south. The two villages, Rhakotis and Pharos, were connected by a causeway called Heptastadion, dividing the harbour into the east and west. The eastern harbour boasted a port with palaces, gardens and government buildings and remained active during both the Ptolemaic and Roman periods.[7] Alexandria remained the Egyptian capital city until the Arab invasion (646 CE) and was one of seventeen cities of the same name founded by Alexander the Great around the world, although it is the only one still standing.

Alexander famously visited the Apis bull at Memphis. For many centuries, this site had been the centre of Egyptian kingship rituals and was a way of Alexander winning over the local Egyptians. The cult of Osiris-Apis was incorporated into Alexandria as the cult of Serapis by Ptolemy III Soter (246–242 BCE) and became the centre of Graeco-Roman worship. Serapis was a solar deity who was closely associated with the Apis bull of Memphis. However, the cult very quickly spread to other Hellenic towns and cities. The Serapeum of Alexandria, based in the Egyptian quarter of the city, Rhakotis, was not only the main cult temple of Serapis but also an art gallery and library and was a great draw for religious visitors, scholars and tourists. There was also an incubation centre where visitors came hoping for a cure for a number of different ailments (see chapter 7).

After visiting the Apis bull in (332 BCE), Alexander the Great took a trip to the Oasis of Siwa to visit the oracle of Amun (see chapter 7) where the god proclaimed him the rightful king of Egypt and a deity. This was another way of ingratiating himself into the local culture and making acceptance of his leadership smoother. From Siwa he travelled to Memphis where he was crowned at the temple of Ptah, as had been traditional for all Egyptian kings for centuries.[8]

## Graeco-Roman Tourists

Alexandria and the wider sites of Egypt attracted a diverse bunch of travellers which included intellectuals, poets, pilgrims, soldiers, politicians and rulers who all had different experiences and expectations of their trips. However, although Egypt was open to travellers, for Alexandrians travel was carefully monitored and Strabo comments they could not leave Alexandria without some dispensations:

> But he could never have left Alexandria without a passport, still less after having stolen the royal property. To set sail on the sly was impossible, as the port and every other exit was kept by a numerous guard, which still exists, as we very well know who have lived in Alexandria for a long time, although it is not so strict since the Romans have had possession, but under the kings the guards were infinitely more alert. (2.3.5)

Egypt had a number of high-profile visitors other than Alexander the Great (330 BCE), including Germanicus (15 BCE), Julius Caesar (47 BCE) and Augustus Caesar (30 BCE). They were all in Egypt for political purposes but still were able to add sightseeing and tourism to their busy schedules.

From the Ptolemaic period onwards Alexandria was a hub of Greek visitors and there were seven Greek embassies in the city which were each responsible for distinct aspects of Greek culture such as games, Eleusinian mysteries and Greek politics. There were inevitably several diplomatic visitors to the city and all were treated to a tour of the best places Egypt had to offer including Alexandria itself, Siwa and the Fayoum.

Many Hellenes arrived to receive counsel from the oracle at Siwa, in the footsteps of Alexander, and others came to witness Ptolemaic religious festivals. Throughout the Ptolemaic period guests from Greece and Rome came to Alexandria to be entertained by the rulers and it is recorded that Ptolemy II Philadelphus (285–246 BCE) entertained diplomats at his palace with his zoo which included horses, dogs, asses, elephants and a massive snake. He was also said to have been crossbreeding pheasants with Numidian birds to create weird and wonderful creatures.[9] Philadelphus had also created the is-olympic Ptolemaia games, in honour of his deified parents Ptolemy I Soter and Berenice. There were viewing stands for visitors including soldiers, artisans and tourists.[10]

Ptolemy VIII Euergertes II (170–163/145–116 BCE) treated the Roman ambassador to a cruise down the Nile and Lucius Memmius in 112 BCE was the earliest recorded Roman tourist. He was taken on a trip to Petesouchos, the sacred crocodile of Arsinoe. His sponsors on the trip even provided him with snacks for the crocodile to further enhance his experience. Lucius then went on to tour the labyrinth of Amenemhat III at Hawara which was in actuality the pyramid structure of Amenemhat III (1929–1895 BCE) rather than a Greek styled labyrinth.[11] Throughout the journey the priests were tasked with making sure there were guest chambers available and landing stages for the boats, as well as numerous gifts along the journey.[12]

> To Asklepiades. Lucius Memmius, a Roman senator, who occupies a position of great dignity and honour, is making the voyage from Alexandria to the Arsinoite nome to see the sights ... take care that at the proper spots the guest-chambers be prepared and the landing places to them be got ready with great care, and that the gifts of hospitality ... be presented to him at the landing place.[13]

Egypt was, in fact, so popular with Greek visitors that Pansanias, a Greek traveller (119–180 CE), glibly commented that Greeks preferred looking at the marvels of foreign countries over their native ones, stating that the treasury of Minyas at Orchomenos and the walls of Tityns at Argos were comparable to the pyramids of Giza.[14]

In the *Odyssey* Homer talks of how the mythical King Manalaeus was one of the earliest visitors to Egypt, who travelled there on the way to Sparta from the Trojan War. Homer's poem became a great motivation for Greek visitors who wanted to travel in the footsteps of the great king.

It is interesting to consider that the reasons people travelled in the first centuries BCE and CE were pretty much the same reasons people travel today regardless of whether they were great leaders or ordinary people. Lucius Memmius, for example, clearly travelled with an interest in the monuments as did Germanicus in 19 CE who was interested in the antiquities of Egypt, whereas Septimus Severus (193–211 CE) was interested in Egyptian religion and especially the animal worship displayed. Greek traveller Cleombrotus, however, travelled, 'out of love of knowledge and a fondness for sight-seeing'. Seneca (4–65 CE) travelled to Egypt for his health and Scipio Aemillianus (185–129 BCE) liked to people watch.[15] Every single reason for travelling in the past is mimicked by travellers today, and just like today, visitors of note were treated very much as VIPs and Germanicus records the collection of corn in order to prepare for his arrival in Thebes.[16]

In addition to the high-status political visitors there were more day-to-day travellers; people who travelled from city to city to visit friends and family, for religious pilgrimage or for business purposes. For example, a letter from a woman called Tarem (second to third century CE), tells how she decided to travel with Persion to reach an unnamed individual who had, 'recovered his mind and spirit'. Tarem then wrote to her father Chairemon, that he should only come as well 'if you find the occasion',[17] demonstrating that travel could be a spontaneous activity.

Travel in the Graeco-Roman period was complex, could take a long time and could be fraught with danger. Fortunately, some of these daily difficulties are recorded in correspondence. Take a woman named Aphrodite, for example, who wrote about her trip to Alexandria when she had her foot trodden on by a horse: 'I was in danger, so that have healed at great expense, and until today I have been out of action'[18] Even without having to stop for rehabilitation, journey times were long. A woman named Isis, who travelled from Philadelphia in the Fayoum to Alexandria in third century CE, wrote to her mother Thermouthion to say she had arrived safely after four days travel – a pace of approximately 75 kilometres per day.[19] Another letter, from Herais to Lucretias (second century CE) asks about travel arrangements. Herais, who was probably in Alexandria, planned to visit her friend in Antinoopolis but, 'if it turns out that you are not going, then write me too so that I won't go uselessly from this place to that'.[20] All of these notes and discussions about travel could have been written at any point in the last 2,000 years, showing the very nature of travel has not changed that much.

## Tourist Season

The tourist season was very much governed by the Nile inundation and the climate, and to a certain degree, this is reflected in the modern tourist season of September through until April when the weather is cooler. In the Graeco-Roman period the Nile flooded from early June starting in the south at Aswan and then about a month later in the Delta and did not recede until October. It was at its lowest levels in April or May. Some travel was facilitated by the flood water, but as it was the hottest time of the year, these months were generally avoided by foreign visitors. However, Egyptian farmers were forced into idleness due to the floods as their lands were under water, and therefore had the opportunity to travel during this time if they had the means to do so. Sadly, most farmers were illiterate meaning there are no records of such trips should they have taken place.

Of course, there is always an outlier, and a strategos (military general) from the Ombite nome visited the Valley of the Kings in the summer and is the only person recorded as doing so.[21] In general, Nubian sites had more visitors in the autumn and winter when the temperature was cooler whereas at the Valley of the Kings tourist season seemed to be from late winter until the Nile flooded in June, and the Colossi of Memnon was also more popular in the winter.[22] Talmis (Kalabsha) in Nubia, had more tourists in May or June which may have been due to the work schedules as visitors were primarily soldiers.[23] Some sites, however, had more visitors during the floods which indicates that they were part of the flood traffic, like landowners and were probably rendered 'unemployed' during those months or the sites were more accessible at this time of year.

However, when people were travelling for festivals or personal reasons that did not fit within the tourist season, they travelled when it was required. Then as now, taking a trip for a festival and meeting with friends had to be planned in advance. One note from Petosiris (third–fourth century CE), discovered in Oxyrhynchus, asks someone named Serenia to come to the festival to celebrate the birthday of the god, but wants to know practical things such as whether she would be travelling by boat or by donkey,[24] indicating the journey may not have been a great distance. Another letter from Serenilla (second century CE) discusses how she had to cancel her plans to travel to Oxyrhynchus in the north for the festival of the Dog Star: 'I was waiting in the south before the festival, and I would be there already, if I had not been bitten by a mad dog, the same day as the rising of the Dog Star.'[25] This could not have been a great omen for the lady and cancelling the trip may have been in her best interests.

## Roman Itineraries

Graeco-Roman visitors followed an itinerary of the most interesting and historical sites to visit. The Egypt the Romans visited was one of ruined monuments, representative of an empire which was once great and would only speak to those who were willing to listen.[26] Strabo records a visit to Heliopolis in the late first century BCE: 'The city is now entirely deserted ... It is said that this place in particular was in ancient times a settlement of priests who studied philosophy and astronomy; but both this organisation and its pursuits have now disappeared.'[27]

Travelling around Egypt was not quick as travellers were still reliant on river boats and donkeys to get from place to place. It could take two weeks for a messenger to travel from Memphis in the north to Aswan in the far south.[28] However, the time did not prevent people from travelling the entire distance from Alexandria to Aswan. Nearchos, for example, travelling in the first century CE wrote to Heliodorus about his expedition to Egypt where he described his route from Aswan to the oracle of Amun at Siwa, stopping in Thebes to visit the tombs in the Valley of the Kings. In what was to become a tradition, he carved his name into the rock to show he had visited the site.

Other itineraries were recorded by later scholars for their historical interest. Take Germanicus' trip which is recorded by Tacitus (56–120 CE). It is said he started in Alexandria, arriving by boat in 19 CE, from where he travelled to Canopus in the Delta, where there was an 'unceasing water carnival' which had drawn a large crowd.[29] He then headed down to the pyramids at Giza, the Fayoum and Memphis. However, the Memphis stop was edited out as the Apis bull would not eat the offering that Germanicus provided. This was thought to be a bad omen. Considering that the bull was likely offered food by every visitor as part of the trip, it could just be that he was not hungry by the time Germanicus arrived. He then travelled south to Thebes stopping at the Colossi of Memnon where he hoped to hear the statues sing as he had learnt in his Homeric education. Then he travelled up the river to Elephantine and Aswan.

Hadrian's visit to Egypt is also carefully recorded and started at Alexandria in August 130 CE. The first thing he did was to visit the tomb of Pompey the Great at the Pelusaic mouth of the Nile, by the foot of Mons Casium, where he apparently placed the head and the body back together. The tomb of Pompey, which is now lost, was a draw for many Greek and Roman visitors. For a long time Pompey's Pillar was rumoured to mark the spot of the tomb and was said to have contained a globe holding the head of Pompey himself; no doubt embellished by tour guides to anyone who listened.

Hadrian then went hunting with Antinous in the desert before heading south on the Nile, founding the city of Antinoopolis in October of that year, near where Antinous was believed to have drowned. The records show that it took two months for preparations to be made for a trip to Thebes,[30] where he arrived at the Colossi of Memnon in November. The first day he arrived the statue did not make the traditional sound at dawn, so he returned the next day. It is suggested that it may have been considered a bad omen for him to have left without having heard it moan.[31] His visit to the statue increased the popular interest and visits to the site as attested by the increased tourism graffiti following his visit. He then started his return journey north stopping at Oxyrhynchus, Tebtunis, Arsinoe, Memphis, Heliopolis and Canopus before arriving back in Alexandria.

At Arsinoe (Crocodopolis), priests entertained tourists by allowing them to feed the sacred crocodiles in the same way priests encouraged visitors to feed the Apis bull. If the crocodile refused the food, it was a bad omen for a tourist. Strabo records how his party fed a crocodile with cake, meat and a honey drink. The crocodile was lying on the bank when the priest approached and held the crocodile's mouth open, throwing in the food that they had brought as an offering.[32] The crocodile then submerged into the lake and swam to the other side, no doubt to prevent it happening again.

Much of the graffiti left by visitors at Egyptian sites seems to have held similar significance to modern photographs taken by tourists 'holding up' the leaning tower of Pisa, for instance, or seemingly putting their fingers on the top of pyramids. Many of the inscriptions start in a rather formulaic way and it is thought these could be phrases repeated by the tour guides. For example, graffiti at the Colossi of Memnon often starts with 'not headless is the son of Dawn, Memnon, since every day at the rising of the sun he prophesies to mortals'. Other graffiti signatures left on various monuments follow the formula 'kissing the ground before the feet of a god or ruler' as a sign of devotion,[33] although in many instances there was no god or ruler named indicating the meaning of the graffiti was closer to a 'so-and-so was here' rather than being truly religious statements. One piece of graffiti in the temple of Mandoulis at Talmis has a visitor, a cavalryman, replacing the name of a god with his horse showing there really was little religious sentiment behind many messages left by Roman tourists.[34] Graffiti was a means for names to be recorded for prosperity in places of note.

KV2 in the Valley of the Kings, belonging to Ramses IV, was the only tomb which could be seen as a religious destination for some visitors. In the fifth century CE, it was used as a Christian chapel, as the increase in Coptic graffiti suggests. What is interesting to note is that some of the Christians who visited this church may have also visited KV9 (FIG 14), which was also open at the time, to see how a pagan site compared, although they did not leave any corresponding graffiti.

Tourist graffiti shows there were some people who signed their names in more than one place, indicating they were seasoned travellers and giving further insight into itineraries. For example, there was one visitor who signed his name in both the tombs in the Valley of the Kings and at the Memnonium at Abydos. Others signed at the Valley of the Kings and the Colossi of Memnon. It seems that most visitors in Thebes travelled to the Colossi of Memnon first at dawn, and then travelled to the Valley of the Kings, as there is graffiti in tombs which mention the Colossi, but none on the Colossi which mention visiting the tombs thus indicating the statues were first on the itinerary.[35]

Other visiting combinations were Colossi of Memnon followed by Philae, Deir el Bahri, Talmis or the pyramids. Of course visitors did not necessarily sign their names at every site they visited so it is likely there were other stops along the way. Another interesting observation is that graffiti also indicates more scholars and intellectuals visited the Valley of the Kings than other sites, with doctors, lawyers and philosophers desperate to visit KV9, which was believed to be the tomb of 'Memnon', based on similarities on the cartouches on the Colossi and the tomb owner Ramses VI (Nebmaatre).[36]

## The First Nile Cruise

As we saw in chapters 3 and 4, the Nile cruise gained traction in the eighteenth century and increased in popularity when Thomas Cook came on the scene in the nineteenth century. However, the very first recorded Nile cruise was in 47 BCE. Greek historian, but Roman citizen, Appian and Roman historian, Suetonius, both record that Cleopatra VII and Julius Caesar embarked on a Nile cruise from the capital in Alexandria to Upper Egypt. It is suggested by some that Cleopatra may have already been pregnant with Caesar's child at this point,[37] although conflicting reports state their son Caesarion was not born until after the death of Caesar, indicating she conceived on a two-year visit to Rome in 46–44 BCE.[38] Nevertheless, this cruise which whilst a pleasure trip, was the ultimate PR stunt with the intention of repairing Cleopatra's relationship with the people of Middle and Lower Egypt following the hostilities caused by her brother/husband Ptolemy XIII killing Pompey and enraging Caesar and by default Rome. Taking Caesar along presented a united front to the people of Egypt showing that Egypt now had the full support of Rome. It was not a low-key affair, and the elaborate royal barge was accompanied by 400 ships crewed by Roman soldiers. We do not have a description of the royal barge, but it was likely to have been an impressive display with elaborately decorated rooms, and dozens of servants to ensure the desires of the queen and Caesar were catered for.

According to Lucan (c. 65 CE) Caesar, like many explorers before and since, was hoping to locate the source of the Nile on this trip:

Despite my strong interest in science, said Caesar to Acoreus, Priest of Isis, nothing would satisfy my intellectual curiosity more fully than to be told what makes the Nile rise. If you can enable me to visit the source, which has been a mystery for so many years, I promise to abandon this civil war.[39]

Unfortunately, Caesar either had bigger ambitions for the extent of the cruise, or he was unsure of the location of the Nile source as the trip likely went no further south than Thebes, perhaps no further south than Memphis.[40] Discovering the source of the Nile was motivated by Strabo who stated that Egypt was renowned for the Nile and its mysterious source which led many travellers to try to know the unknowable.

Even before Cleopatra VII, such a trip from Alexandria down the Nile was not unusual for royalty and the majority of the Ptolemaic kings had toured the country in order to perform religious rites and to commission additions to temples from Alexandria down to Elephantine. Ptolemy VI (180–164/163–145 BCE), for example, visited the Serapeum at Memphis, oversaw the installation of the Buchis bull in Thebes (as did Cleopatra VII) and then continued his journey to Philae to participate in a festival which took place at the end of the flood season.

Ptolemy IX Soter II (116–80 BCE) also made a trip from Alexandria to Elephantine in order to throw gold and silver offerings into the mouth of a cave believed to be the source of the Nile. Despite this history of travelling the full length of the Nile, it suddenly became internationally more interesting when it was an Egyptian queen and a Roman emperor making the journey. The imperial trips down the Nile continued throughout the Roman period, including when Julia Balbilla, accompanied the empress Sabina on a tour in 130 CE, and left a poem on the left leg of one of the Colossi of Memnon:

Yesterday Memnon was silent as he received the wife,
In order that the fair Sabina would return here,
For the exquisite beauty of our queen pleases you.
At her arrival you uttered your divine cry,
Lest our Emperor be angry with you, for too long a time in your boldness
You constrained his holy and legitimate wife.
And Memnon terrified of the power of great Hadrian
Suddenly cried out, and she, hearing, was pleased.[41]

The second recorded cruise down the Nile had more of a heavy-handed military approach than Cleopatra's and Caesar's, when his successor Octavian

Caesar (63–14 CE) travelled from Alexandria to the south, to inspect canals and irrigation channels to make sure they would be able to provide Rome with the agricultural yield he had promised. Rather than visiting the temples and monuments with interest as Cleopatra and Caesar had done, Octavian refused to visit the Apis bull of Memphis as he wanted to suppress such animal cults and claimed he would not pay homage to cattle.[42] However, the priests of Apis were clearly expecting him and they had already commissioned a relief depicting him performing the rites to the Egyptian gods.[43] According to Egyptian religious ideology, to depict something taking place, whether food being offered or a ruler performing rites, is for it to take place for eternity.

Octavian also visited the tomb of Alexander the Great but refused to visit any Ptolemaic tombs preferring to keep his distance from the Egyptian rulers. Suetonius records this visit:

> he had the sarcophagus containing Alexander the Great's mummy removed from its shrine and, after a long look at its features, showed his veneration by crowning the head with a golden diadem and strewing flowers on the trunk. When asked 'would you now like to visit the mausoleum of the Ptolemies?' He replied: 'I came to see a king, not a row of corpses.'[44]

Dio, in his record, adds that when Octavian touched the figure of Alexander, he accidently snapped off the nose.

Alexander the Great was admired by the Romans and his tomb was a principal tourist attraction in Alexandria. Alexander's body had been moved regularly for two centuries until the Christian era. He had died in 323 BCE in Babylon, and Egyptian embalmers were called to the city, as the best in the business. He was expected to be lain in a golden coffin filled with aromatic spices and a purple cover laid over the coffin displaying his shield and armour. The coffin was to be carried through the streets on an elaborately decorated chariot depicting scenes from his life. In 321 BCE the procession left Babylon on the way to the burial ground at Aegae (modern Vergina in Greece). When it reached Damascus, Ptolemy, son of Lagos the governor of Egypt, intercepted the procession and announced that Alexander would be buried in Egypt. Ptolemy claimed on his death bed that Alexander had expressed the wish to be buried at the temple of Jupiter-Ammon at Siwa. No one else had heard this desire and as Siwa was remote and Ptolemy believed the tomb would be valuable as a draw for visitors he decided to bury him elsewhere.

Alexander was first buried in a temonos in a sacred precinct in Memphis and during the reign of Ptolemy II (285–246 BCE) was transferred to Alexandria. It is thought that an alabaster box in the Catholic cemetery of Terra Santa may have formed part of this first Alexandrian tomb. He remained here for 300 years in the gold coffin, until Ptolemy X (110–88 BCE) plundered the gold and replaced it with an alabaster coffin.[45] After the visit of the emperor Caracalla (188–217 CE) the tomb of Alexander appears to have been lost and to this day archaeologists, scholars and enthusiasts still hope to one day find it.

*Alexandria – Mouseion and Library*
Most travellers started in Alexandria and in the Graeco-Roman period it was known for the library and mouseion as centres of learning. The mouseion housed the famous library, museum and was an important seat of learning with departments including poetry, rhetoric, philosophy and medicine. The concept behind the library was that of an Athenian politician, Demetrius of Phaleron, who fled to Egypt from Athens and became the advisor of Ptolemy I, and the mouseion was built between 305 and 282 BCE. It was centrally based near the harbour and the royal palaces. Strabo described it:

> And the city contains most beautiful public precincts and also the royal palaces, which constitute one-fourth or even one-third of the whole circuit of the city; for just as each of the kings, from love of splendour, was wont to add some adornment to the public monuments, so also he would invest himself at his own expense with a residence, in addition to those already built, so that now, to quote the words of the poet 'there is building upon building'. All, however, relate to one another and the harbour, even those that lie outside the harbour. The museum is also a part of the royal palaces; it has a public walk, an exedra with seats, and a large house, in which is the common mess-hall of the men of learning who share the museum. This group of men not only hold property in common, but also have a priest in charge of the museum, who formerly was appointed by the kings, but is now appointed by Caesar.[46]

The Library was stated by Strabo to have been modelled on that of Aristotle who was believed to be the first person to have owned a library. Aristotle's library was said to have been deposited in Alexandria, which led medieval scholars to the conclusion that Aristotle taught here.

The library was more than a book depository and Strabo describes the shared dining room and covered walkways with seating which was filled with scholars debating. Eusebius records that Ptolemy I: 'had the ambition to equip the library, established by him in Alexandria, with the writings of all men, as far as they are worth serious attention'.[47] Boasting a copy of every book written, the library of Alexandria had texts in numerous languages, with the most common being Greek and Egyptian. There were also specially commissioned Greek translations of works originally written in foreign languages as well as the works of great philosophers, mathematicians and physicians. The Ptolemaic kings sent out parties to every ship arriving in the harbour searching for books which were meant to be taken to the library and then copied, with the original being returned to the owner.

However, Ptolemy III (246–222 BCE) apparently stole the original manuscripts of Aeschylus, Sophocles and Euripides from the Athenian archives, under the pretence of borrowing the manuscripts to be copied. He kept the originals and returned the copies. Other books were bought above board, with the most influential booksellers being in Athens and Rhodes. Through this aggressive collecting technique, the library was recorded as having 400,000 mixed scrolls (more than one text on each scroll) and 90,000 single-text scrolls. Another sub-library was constructed at the Serapeum which held another 42,800 scrolls which included the work of Plato, Aristotle and Xenophon, and it is possible that Herodotus prepared his guidebook, *The Histories*, at the mouseion. At its height, Aphthonius, (second half of fourth century CE) described the library:

> on the inner side of the colonnade, were built chambers, some of which served as book-store and were open to those who devoted their life to the cause of learning ... Other rooms were set up for the worship of old gods. The colonnades were roofed and the roof was made of gold and the capitals of the columns were made of bronze overlaid with gold.[48]

The library had an international reputation as being the home of many secrets, and Pliny the Elder (d. 79 CE) who was employed at the Arsinoeion, which was built to honour Arsinoe II, talks of a statue of the queen within the shrine which was suspended using magnetism:

> The architect Dionochares had begun to use loadstone for constructing the vaulting in the temple of Arsinoe at Alexandria, so that the iron statue contained in it might have the appearance of being suspended in mid-air; but the project was interrupted by his own death and that of King Ptolemy who had ordered the work to be done in honour of his sister.[49]

The secrets held in the library were all destroyed when it burnt to the ground in 48–47 BCE so whether this was possible will remain unknown.

The burning of the library was a tragic accident which rocked the ancient world, with 700,000 volumes destroyed. Julius Caesar records in his *Civil War* that he burned the vessels in the Alexandrian harbour as a means of protecting himself against Pompey as he attempted to capture Alexandria. The fire then spread to the warehouses and the cargo on the wharf, which according to Galen also included more books for the library.[50] It then spread to the library itself. Some believe that not all the library was burnt at this time and Marcus Antonius did his best to start replenishing the losses by gifting Cleopatra VII (51–30 BCE) the library of Pergamum which amounted to 200,000 volumes,[51] and scholars from the mouseion procured books for the library. In 12 BCE Augustus presented a new library indicating that replenishing it was important to him too.[52] Unfortunately this library was also not to survive, and evidence suggests that by the end of the fourth century CE the newly replenished library had also burned to the ground.[53] Epiphanius of Salamis (c. 403 CE) records that the area where the mouseion once stood was now abandoned.

At its height, the library of Alexandria was considered to be a site worthy of tourism and it is listed by Greek poet Herondas as an attraction to draw people to Alexandria. The library was more than a depository of books, it was a centre of learning where scholars and philosophers from all over the known world came to learn, and to study with the top thinkers of the day. Plutarch stated that only the wisest Greeks travelled to Egypt in order to improve their knowledge. The library was also a place where people of power came for refuge or for political dealings. Lectures were held within the exedra (seating areas), which were open to scholars, and it is even thought that Julius Caesar attended lectures here as a means of showing he was no threat to the Alexandrian people.[54] It was essentially the centre of the intellectual world during the Graeco-Roman period. Evidence would also suggest that visiting scholars were offered patronage, enjoyed a tax-free status within the city and had the right to live in the royal quarter in Alexandria.[55]

While Alexandria was clearly an important place for scholars to visit, there were numerous other sites which attracted tourists.

*Naucrautis*
Naucratis, situated on the Canopic branch of the Nile in the Delta was an important site for Greek visitors both as a trading centre from at least the twenty-sixth dynasty (664–525 BCE) as well as an important Greek religious centre. Visitors often went there for religious pilgrimage, and instead of leaving graffiti as a commemoration of their visit, they dedicated vases which were inscribed with their names as well as the names of the Greek gods worshipped at the site. These included Aphrodite, Hera and Apollo, and there was also an

Egyptian temple dedicated to Amun. Two vase fragments have been found which bear the name Herodotus, and the date and style of the vase suggest that they could have been dedicated by the man himself.[56]

*Pyramids of Giza*
Many of the monuments in Egypt, such as the pyramids, for example, were already 2,000 years old before Cleopatra and Julius Caesar saw them in 47 BCE, and they were considered awe-inspiring and noteworthy throughout the ancient world. Two of the three Graeco-Roman inscriptions at the top of the pyramid show some travellers came simply to admire the monuments, exactly as travellers have been doing ever since. However, not everyone liked the pyramids and Pliny the Elder (50 CE) seemed to think they were somewhat ostentatious.

> We will mention also cursorily the Pyramids, which are in the same country of Egypt, – that idle and foolish exhibition of royal wealth. For the cause by most assigned for their construction is an intention on the part of those kings to exhaust their treasures, rather than leave them to successors or plotting rivals, or to keep the people from idleness. Great was the vanity of those individuals on this point.[57]

The smallest of the pyramids built alongside the Great Pyramid was rumoured (started by Herodotus) to be the tomb paid for through prostitution as a tool to show how evil the king was:

> But no crime was too great for Cheops: when he was short of money, he sent his daughter to a bawdy house with instructions to charge a certain sum – they did not tell me how much. This she did, adding to it a further transaction of her own: for with the intention of leaving something to be remembered by after her death, she asked each of her customers to give her a block of stone, and of these stones (the story goes) was built the middle pyramid of the three which stand in front of the Great Pyramid.[58]

One wonders if Herodotus believed this himself, as his interjection of 'as the story goes' would indicate otherwise. He seems to want to make it very clear that this is what he was told and not necessarily fact. This pyramid does, in fact, belong to one of the queens of Khufu although Egyptologists cannot agree on the exact queen, but it is commonly stated to belong to Meretites.

During the Roman period the Sphinx was once more buried in sand and Tiberius Claudius Balbillus, the Egypto-Syrian prefect of Egypt in 55 CE sent a petition to Emperor Nero requesting that the sand be cleared away, possibly the first time since Thutmosis IV (1419–1386 BCE) uncovered the monument which he recorded on the Dream Stela and erected at the foot of the Sphinx (chapter 8).[59] Due to it being buried, the Sphinx was not written about until Pliny the Elder (d. 79 CE). Following this, the Sphinx became an additional stop on the pyramid tour. It was referred to in Roman literature as a 'divine spectacle' and 'frightening vision',[60] showing there was no religious interest or knowledge, and they were literally visiting to see the wonder of the monument.

### *Tell el Amarna*

The city of Tell el Amarna in Middle Egypt was largely quarried during the pharaonic period following its abandonment after the death of Akhenaten. However, the Greek graffiti at the site shows that for some it was an attractive if unusual site to visit.

Tombs of officials seem to have been accessible at this time and much of the graffiti dates to the second century BCE. The names recorded suggest they were Ptolemaic mercenaries of Thracian descent. One of the inscriptions of Spartakos, a runner in Alexander's games in Memphis, suggests that Alexander the Great himself may even have travelled as far south as Amarna.[61] As it was off the tourist trail and was not mentioned in the guidebooks of the time, it was probably quite hidden, with no guides or easy access. But at least one tourist, Catullinus, believed it was a site worth visiting.[62] When visiting the site he left graffiti in Greek to the right of the doorway of the tomb of Ahmose (EA3) to say how he marvelled at the art he saw there.

### *Abydos*

Although for the most part, Graeco-Roman travellers to Egypt were tourists in the true sense of the word, there were also a number of people who travelled as pilgrims to worship particular deities at specific sites such as Isis at Philae, or to leave sacrifices as Herodotus tells us happened at Bubastis. Many were non-Egyptians, or at least left their commemorative graffiti in non-Egyptian languages. Visitors appeared to be from Greece, Thrace and Asia Minor. However, there were some Egyptians who were visiting the site on pilgrimage with many coming from Ptolemais, Thebaid, Oxyrhynchus, Kunopolites Nomos, Cusae and Antinoopolis. In some instances it is difficult to identify the country of origin, as the visitor claims to be from one place but lives in Egypt. Poluarates, for example, describes himself as 'Cyrenean, but now the land of Egypt holds him', making it difficult to identify how far they travelled to get to the site. It is clear, however, that pilgrims tried to make regular visits to the site and one Thracian soldier from Cusae in Middle Egypt records that he visited Abydos at least eight times on pilgrimage (based on appearances of his name).⁶³

Abydos was a particularly popular place for pilgrimages during this period, especially the Memnonium (FIG 15), which was what the Greeks called the mortuary temple of Sety I. During the Roman period it was approached as a temple of Osiris, then Serapis (a combination of Osiris, Apis and Zeus) and then Bes, the dwarf god of childbirth. Many visited the temple of Serapis in order to be healed from an illness, and then to thank Serapis for healing them, as well as for dream oracles (chapter 7). Many of the pilgrims visited twice and Sphex, for example announced that he was visiting the oracle for a second time, 'this time healthy'.⁶⁴ Many visitors to Bes, on the other hand, appear to be athletes who had won at the Olympic Games or the Pythian Games to either ask for his advice or thank him for their wins.

It is thought that the pilgrims at this time may have gravitated towards this part of the sacred complex as the only place which was open to them at the time.⁶⁵ However, over time the areas open to pilgrims changed and this is indicated by the location of the graffiti. For example in the early Graeco-Roman period the graffiti was located in the Memnonion, on the staircase near the king list, and includes Greek, Phoenician and Aramaic inscriptions. As time progressed, pilgrims had access to the more sacred areas of the temple including the Hypostyle Hall and the chapels of Isis, Osiris and Horus.⁶⁶ Some visitors came for a combination of religious piousness as well as intellectual tourism. Philokles left some graffiti in the Chapel of Isis which expressed he was visiting to worship Serapis.⁶⁷ It is thought that in the second century BCE the Memnonion may have been used as a military barracks as a means of quelling revolts which were happening in the area.⁶⁸

The site of Abydos had been continually active from the Old Kingdom (2686–2181 BCE) until 359 CE when Constantine closed it down following a number of questions posed to the oracle about succession (chapter 7) which may have been written on the walls for all the visitors to see. However, there was still a following of pilgrims to the site until the Christians closed it in the second half of the fifth century.

*Colossi of Memnon*

The Colossi of Memnon once stood at the pylon entrance to an impressive temple built by Amenhotep III (1386–1349 BCE), but by the Graeco-Roman period all that was left were two colossal statues of the seated king (FIG 16). Memnon was known to the Greek tourists from Homer, portrayed as the king of the Ethiopians in the *Odyssey*, whose mother, Eos (Dawn), cried when he was killed by Achilles. The statues were connect to Homer's Memnon in approximately 20 CE and were a means of associating the statues with the heroic past.

The connection was made when following an earthquake shortly after the Roman conquest (30 BCE), half of one of the statues fell to ground, and as the damaged stones started to warm up in the dawn sunlight, they emitted a low keening sound which Greek visitors equated with Memnon crying for his mother. The spectacle happened nearly every dawn attracting numerous tourists, many of which carved their names on the legs of the statue to commemorate their visit.

Many of those who left graffiti identified as artists and poets and were there because of the Homeric associations. One particularly moving piece of graffiti was written by a woman called Treboulla, who missed her mother (who was perhaps dead) and was sad that she also would not be able to hear Memnon.[69] She may have been a poet as she left three verses at the site, although she does not refer to herself in this way.[70]

Evidence suggests that a mini-tourist industry had arisen around the statues with guides and stonecutters on hand. As some of the graffiti were carved so deeply into the stone they could only have been carried out by a professional stonecutter and would have taken some time to execute. The earliest graffito is dated to 65 CE and the latest was 196 CE by the prefect of Egypt. This site saw a regular stream of visitors and guides were able to provide extra information to them, as Julia Balbilla added to her inscription an alternative name to Memnon, that of Amenhotep which she had picked up on the trip from a guide.[71] It is thought these guides were possibly priests as they were the only group at this time who were able to read hieroglyphs. Graffiti shows visitors, including seven prefects of Egypt as well as most of Hadrian's retinue, intellectuals and royal visitors, like Julia Balbilla and the empress Sabina on a tour in 130 CE. The majority of the graffiti at the Colossi of Memnon was left by soldiers and administrators, whereas the Valley of the Kings was the opposite, with more intellectuals, doctors, philosophers and teachers although no royals are recorded as visiting.

Could this be due to the location and ease of access? The Colossi of Memnon was a minimum effort attraction, whereas to visit the tombs required scrabbling down little holes into hot, dirty and tight spaces and therefore took a lot more effort. Perhaps the effort was proportional to the interest in the monument. And there was definitely interest, with the centurion Lucius Tanicius visiting the statues thirteen times between 80–82 CE and every time he recorded the time and date that the statue sang for him.[72]

But the Colossi of Memnon as a tourist attraction was not to last. The statue was repaired under Septimius Severus in 205 CE and was silenced forever. Severus wrote the last piece of graffiti on the statue, and it can be assumed that it was not considered such a worthy place to visit as it no longer cried at dawn.

*Valley of the Kings*
Strabo's *Geography* refers to the Valley of the Kings, showing that some of the tombs were accessible at this time: 'Above the Memnonium [the Ramesseum], in caves, are the tombs of kings, which are hewn from stone, are about forty in number, are marvellously constructed and a sight worth seeing.'[73] However, they had been available much earlier, with the earliest graffiti at the site dated to the reign of Ptolemy II (285–246 BCE) and continuing for another 600 years. The earliest record about the Valley of the Kings, after the workmen who built it, comes from Diodorus (60–30 BCE) who wrote:

> The priests said that in their records they find forty-seven tombs of kings, but down to the time of Ptolemy, son of Lagos, they say only seventeen remained, most of which had been destroyed by the time that we visited these regions.[74]

To reach the tombs in the Valley of the Kings, visitors needed to hike through the Theban hills, and this seemed to have been the case from the Ptolemaic period through to the fifth century CE. Both men and women made the journey and one woman, Latina, wrote on the cliffs that she was visiting from Rome. Some inscriptions show that some hikers believed the mountain to be sacred and were therefore possibly local Egyptians who referred to 'the great gods in the holy rock' and considered it a 'sacred place'.[75]

Ten of the tombs from the Valley of the Kings contain Graeco-Roman, Demotic, Phoenician and Cypriot graffiti, indicating they were easily accessible and well-visited between 278 BCE and the fourth century CE. These are:

- KV1 – Ramses VII
- KV2 – Ramses IV
- KV4 – Ramses XI
- KV6 – Ramses IX
- KV7 – Ramses II
- KV8 – Merenptah
- KV9 – Ramses V and Ramses VI
- KV10 – Amenmesse
- KV11 – Ramses III
- KV15 – Sety II

KV9, the tomb of Ramses V and Ramses VI, was a particularly popular tomb as it was believed to have belonged to Memnon and contains more than 1,000 pieces of commemorative graffiti, half of all found in the Valley of the Kings. The Greek visitors believed this identification with Memnon was supported by the hieroglyphs which they could not read, but which local guides at the site helped translate. The connection was likely due to Ramses VI having the same pre-nomen as Amenhotep III (Nebmaatra), connecting the tomb with the Colossi of Memnon. You can imagine this similarity being pointed out by over-eager guides who accompanied the tourists to the two sites.

The graffiti at the Valley of the Kings also tells us how far people travelled to visit. For example there were Massiolotes, an Olynthian, a Byzantian, twenty-three tourists from the Greek islands, forty-two from Asia-Minor and thirty-one from the Greek mainland.[76] Unfortunately, the graffiti does not really tell us how many people in total visited the Valley of the Kings in the Graeco-Roman period, but it can give us an insight into the visitors' experience whilst they were there. Those who were predisposed to carving their names tended to do so in all the places they visited, and this shows that individuals named Iasios and Synesios, for example, went into six of the available tombs.

At the time, the tombs were all filled with debris – some left by the original builders, and some caused by flash floods throughout the Valley, meaning that to enter required shuffling through tunnels on their stomachs and once in the decorated chambers, often only the top half of the walls were accessible.

Despite this difficulty tombs were still popular, even though visitors had no real understanding of what they were looking at and who had built them. They simply visited the tombs for the beauty and for the wonderment of the structures. One graffito states, 'I looked, then I investigated, and finally marvelled [at the site] after I arrived,' and Ouranios the Cynic admired the engineering of the tomb and the beauty of the artwork.[77] One well-travelled visitor, Antonius, compared the beauty of the tombs with those found in Rome, where he had apparently spent a great deal of time.[78] However, not everyone enjoyed the tombs as one individual commented 'I, Epiphanius, visited but admired nothing but the stone.'[79]

Most of the records are of male visitors, although women were known to visit, too. For example, a primipilaris (centurion of the first cohort of a Roman legion) named Januarius and his daughter Januarina visited the tombs and 'marvelled at the place'.[80]

It is clear from the graffiti that for the vast majority, they were not visiting as part of a religious pilgrimage – but simply for the historical significance and engineering brilliance. None of the inscriptions left behind make any reference to deities (either depicted in the tombs or not) and those at the

Colossi of Memnon, for instance, have a scholarly slant rather than a religious one, making it clear visitors were there to hear the statue cry at dawn.

There has also a vibrant community living at the Valley. Following the work of the Emperor Theodosius in 391 CE when he banned all pagan cults and closed all temples, instigating a mass conversion to Christianity, a community of Christian hermits lived on the west bank of the Nile, some in the tombs of the Valley of the Kings. For example, at the rear of KV3 (unidentified son of Ramses III) a small chapel had been built on the tiled floor complete with Corinthian columns and a smaller church was constructed at the entrance to the tomb of Ramses IV (KV2). There was also evidence of residential households complete with cooking vessels, bread ovens and pottery with inscriptions.[81] The tomb of Ramses XI (KV4) had been used by the Christian inhabitants as a stable and a kitchen and archaeologists found straw in a manger within the tomb which fed the horses. The shaft in the front of the burial chamber had been filled in and the rear had been used as a rubbish dump which even included the bodies of a number of unwanted puppies. It is clear there was an active community here, although not particularly compassionate when it came to animals.

As J. Romer explains,

> Among the lonely desert cliffs above the Nile Valley, the hermits led the simplest of lives. They rejected material possessions, suffered appalling hardships and left a body of writing remarkable for their faith and humility. In their desert solitudes many of them founded strange forms of Christianity that once again attracted interest.[82]

In these tombs people also carved Christian slogans and prayers alongside the images of the Egyptian gods in the tombs:[83] 'I beseech thee, Jesus Christ, my Lord, suffer me not to follow after my desire: let not my thought have domination over me: let me not die in my sins, but accept thy servant for good.'[84]

The Valley was clearly a site of pilgrimage, and for some it was like visiting the underworld itself, an area inhabited by gods and demons, although as Christians they did not worship here. The latest piece of graffiti is dated to 537 CE and was left by Count Orion who was the governor of Upper Egypt. When the Arab invasion happened this community was either coerced into changing to the new religion or they were executed, leaving the Valley of Kings once more abandoned.

### Nubian Temples

Many Graeco-Roman visitors also continued southwards from Thebes to Aswan, and Philae was a popular site for different groups of travellers. One group was Egyptian artisans with many claiming to be sculptors, painters, dancers and athletes, all roles that may have necessitated travel in their day-to-day activities and their work.[85] The second group during the Roman period were high ranking foreigners, which may have been due to the near-by location of a garrison on the border of Egypt and Nubia. Other visitors were religious pilgrims who came to honour Isis and be healed by her, as well as Nubian Christians visiting her temple. Isis as a goddess was worshipped beyond Egypt and was a draw for pilgrims until the sixth century CE.

Many of the visitors stated that they were visiting from Kush or Meroe just south of the borders of Egypt, and this even included the Hellenised ambassador of the queen of Nubia, who signed his name at Dakka temple rather than Philae.[86]

One of the attractions at Philae was the white-water boat rides which Strabo talks about as something worthy of seeing and a visitor mentions it in his graffito at the site. Strabo records arriving here:

> We went to Philae from Syenê by wagon through an exceedingly level plain – a distance all told of about one hundred stadia [15 kms]. Along the whole road on either side one could see in many places a stone like our Hermae; it was huge, round, quite smooth, nearly sphere-shaped, and consisted of the black, hard stone from which mortars are made – a smaller stone lying on a larger, and on that stone again another. Sometimes, however, it was only a single stone; and the largest was in diameter no less than twelve feet [3.6 m], though one and all were larger than half this measure. We crossed to the island on a pacton. The pacton is a small boat constructed of withes, so that it resembles woven-work; and though standing in water or seated on small boards, we crossed easily, being afraid without cause, for there is no danger unless the ferry-boat is over-laden. (Book 8: 50)

The oldest Latin inscription at Philae (and indeed Egypt) dates to 26 August 116 BCE, and was left by Actius and Varaeus, with the latter's contribution being interrupted before he could finish it. It is not stated why they were there, but it was within a month of the death of Ptolemy VIII Euergetes II (170–116 BCE) and could have been a reconnaissance mission. Regardless of the official reason for being in the area, it is likely they came to the temple as tourists and left the graffiti as a commemoration of the visit or even as a form of reverence to the gods.[87] The end of August was also an important time at Philae as there were festivities to celebrate the high point of the Nile's inundation. Aswan was believed to have been the location where the Nile first rose and although the date does not coincide precisely with the festival, they could have been in the area to participate in it.

Philae remained an important pilgrimage site for about 700 years from the Ptolemaic period through to fifth century CE when the temple closed. Isis had international appeal, her temples were visited by Egyptians, but also by Greek and later Roman visitors as well as Ethiopians from the south which is attested by some graffiti at the site naming the Ethiopian king Yesbokheamani of the third century CE.[88]

It is thought the evidence of Ethiopian pilgrimage to Phliae may have been for one of two reasons: for religious or for political purposes in order to show their presence in the area. There is some ambiguity as to the roles Ethiopian rulers had in the area, and it is thought they may have had a hand in governance in the form of administrators alongside the Romans. The earliest Ethiopian visitors were between the first century BCE and first century CE. Graffiti was left by Egyptians who record: 'We saw on the Nile, fast-travelling ships which brought the admirable temples of the Ethiopians to our land, wheat-bearing worthy to look at.'[89] Seeing such an unusual and impressive sight was fascinating for the Egyptians who just happened to be at the temple at the same time. It is thought this coincidence may have been due to it being a festival date when people were converging here.

However, the journeys of the Ethiopian visitors did not start and end with Philae but also covered other temples in the area. For example, on Isis' birthday on Thoth 4 (14 September) a boat went from Philae to Dakka to celebrate the yq Festival (third century CE). This festival was also known as the Festival of Entry or Festival of Dedication and was characterised by a journey between Philae and Bigeh (near the first cataract). It would have been an opportunity for Ethiopian royalty to show their wealth and power.

Graffiti at Philae is written in various languages from Greek and Latin, to Demotic and Meroitic. There were probably more Egyptian visitors throughout the earlier years of the history of the temple than is attested by the graffiti, but literacy was low, and they may not have left records of their visits. Many of the Egyptian (Demotic) inscriptions were in the mammisi area of the temple, which was associated with birth and fertility.

Graffiti shows there were different types of visitors; there were those where the 'arriving' at the site was the important part of their trip, whereas others came specifically to offer adoration to the deity on behalf of themselves or friends and relatives who were unable to make the journey. One religious pilgrim, Pathores, was very pleased with his visit and prayed that he would be able to return annually to worship the goddess.[90]

The pilgrims who came here to make adoration to a deity (*proskunemata*) started arriving in the first century BCE and it was at this time, that the increased numbers of visitors encouraged a temple redesign, creating an area in front of the temple which could be used to receive pilgrims. However, by the fifth century there were no out-of-bounds areas and graffiti has been found on the roof of the shrine of Osiris and in the Christian period, inscriptions are within the pronaos which was originally closed off to everyone but the king and the priests.

A Merotic inscription of Wayekiye (229 CE) a political figure records the visit and his view of Isis as a helper:

> I worship before thee, bending the arm unto thee, my mouth filled with praise unstinted. I kiss the ground to thee, adoring thy dignity, thine awe in my belly. Hear my prayers thou mistress of the lands, though mightiest of all the gods. Come thou into my prayers with thy face of graciousness which the gods rejoice to see; hear my entreaties, my mistress, O Isis! And grant me strength to equal that which my elder brethren have done and grant me favour and love and respect in the presence of kings. I am thy servant, my heart is right loyal, thou that hearest the entreaties of them that are far off [even as] them that are nigh to thee, that protectest one against a multitude, putting thine enemy under thy favoured one.[91]

Pilgrimage to Philae was more popular in January, February and March although there was a constant stream of visitors throughout the year although surprisingly there did not seem to be a spike in visitors for the Khoiak festival which fell in December.

Visitors to Philae ranged from Roman governors such as Cornelius Gallus (70–26 BCE) to personnel from the cult of Serapis. There were also several ordinary people from scribes to mimes, to a litter bearer. Three visitors, Serenus, Felix and Apollonius had visited in 191 CE from Alexandria, travelling a distance of 1,069 kilometres and they may have visited other temples in the area.

The site of Talmis (modern Kalabsha) saw a number of visitors to the temple of Mandoulis, a Nubian solar god, who had an oracle on this site (see chapter 7). He was associated with the Greek god Apollo, the oracle giver. The majority of visitors to the site were military men or in administration indicating they were working in the area.[92] Other travellers wrote poetry about their experience and invoked the muses and Homer, using similar language to that found on the Colossi of Memnon, indicating for some it was an intellectual rather than a religious excursion.[93]

The temple at Kalabsha was on the site of one originally built by Amenhotep II (1453–1419 BCE) and it was in use until 250 CE when the Romans left the area. The Blemmyes then took control of the area and continued to use the temple for the deity Mararouk, an incarnation of Mandoulis.[94] It was converted into a Christian church in 500 CE.

Some of the earliest examples of Greek text can be found at Abu Simbel temple. Although Greek pre-dominates there is also Semitic and Carian graffiti which dates to Psammeticus (664–610 BCE) and the later Ptolemaic period (305–30 BCE). We do not know why these people were at the site and it is presumed they were in the military or on bird or elephant hunting expeditions. The lack of Egyptian graffiti could be down to literacy levels, or it could be an outsiders' lack of respect for Egyptian monuments which Egyptians still reverenced and refrained from defacing.[95] There are no Roman inscriptions here which could indicate it was too remote for most visitors and unless there were soldiers in the area the site was left alone.

This period of tourism in Egypt was at a time of great change, with Alexander and then the Romans conquering Egypt which saw a shift in culture, a merging of Greek, Roman and Egyptian religions and the introduction of Christianity. People's motivations for visiting Egypt were varied but it is interesting to note that many had similar purposes regardless of their origin. This is made clear in the visits to oracles, which traversed cultures and gives an insight into the fears and concerns both Egyptians and non-Egyptians had.

Chapter 7

# Visiting the Oracle

A very specific form of pilgrimage throughout the pharaonic and the Graeco-Roman periods was the oracle. This provided an opportunity for the public to converse directly with a deity (sometimes through a mediator) and to get help with day-to-day concerns, health or fertility. Whilst the oracle increased in popularity in the Graeco-Roman period and continued in some form into the Christian tradition (chapter 4), it had been a longstanding tradition in Egypt which extended back to the Old Kingdom (2686–2181 BCE).

There were varying forms of oracle available, although the premise was always the same: the petitioner asked the god a question, and the god answered in one form or another.

## Processional Oracles

One of the most common forms of oracle, particularly during the New Kingdom (1570–1070 BCE) was the processional oracle, which was a one-to-many approach. There were frequent processions in ancient Egypt (as many as ten a month in some places) where a deity was carried by a number of priests in a sacred barque through the streets. The streets were lined by the general populace, many of whom had travelled specifically to be there to catch a glimpse of the holy barque and be in the deity's presence.

However, the studies of the oracles at Deir el Medina, the New Kingdom town on the west bank at Luxor which housed the workmen who built the Valley of the Kings, showed that oracular processions did not always coincide with religious festivals, or scheduled processions dedicated to Amenhotep I or in fact any other deity worshiped in the town.[1] This indicates the oracle, at least at Deir el Medina, may have been organised as a procession in its own right. It is not exactly clear who had the authority to organise an oracular procession but a twentieth-century papyrus (1185–1070 BCE) records how one petitioner at Deir el Medina wrote to a god and requested that he come out in a procession to answer his questions, indicating the public could potentially make the request.[2]

During these processions, people called out questions to the deity – generally those with a simple yes or no answer.[3] Then the deity within the barque compelled the priests carrying it to move in a particular way; for example, to move forward for yes, and back for no, therefore answering the petitioner's question.

There were no limits to the type of questions that could be posed to the god – from those about health, to the wellbeing of relatives in far-flung places, from the Nile inundation to work related queries as described on a papyrus in Brooklyn:

> In year 14 of Psamtik 1, on the fifth day of the first month of Shomu [4 October, 651 BCE], the occasion being the festival of the new lunar month, there was a procession of the August God, Lord of All the Gods Amun-Re, King of the Gods.
>
> His shrine was borne by 20 priests out onto the Floor of Silver in order to go around the temple. Before the shrine were several more important priests, including the First Prophet or High Priest, the Third Prophet and the Chief Lector Priest with the venerable Montemhet, Fourth Prophet of Amun and overseer of Upper Egypt, heading them and offering incense.
>
> In its course the procession arrived at the … Hall of Review, where the ordinary people awaited it. It was here that Pemou, son of Harsiese, son of Peftjau, stood and announced to the great god his presence and besought of him an oracle in favour of his father Harsiese, who wished to leave the service of Amun for that of Montu-Re-Horakhty.
>
> This petition the great god was pleased to grant and he signified his approval by responding to the voice of Pemou and advancing.[4]

The god seemed to show his answer by moving the procession forward ('advancing') to demonstrate the affirmative. This was obviously considered such an important announcement by the oracle that Pemou decided to record it on a long papyrus (more than five metres long) with a large vignette showing the scene which is more than one metre wide. Additionally, all forty-nine of the priests who witnessed it also wrote their accounts on the papyrus – some taking up to five days to record the event even though they all pretty much said the same thing. Pemou depicts himself and his father in the procession, as smaller than the priests to show their inferior position.

Many people appealed to the god in the oracle to identify a thief or locate stolen items. A papyrus in the British Museum (BM 10335), for example, describes such a situation where a man whose job it was to take care of the sacred garments of the god noticed that five tunics of coloured cloth were missing. He appealed to Amun to identify the culprit. He read out the names of potential thieves in the procession and a local farmer was acknowledged by the god to be the thief.

The farmer, Pethauemdiamun, objected and asked to appeal to another oracle in the Theban area, the oracle of Amun of Te-Shenyt, his own local deity – the 'second opinion' gave the same verdict so he then appealed to Amun of Bukenen, who also gave the same response: 'the god nodded very greatly, saying: "it is he who took them." And he took him and inflicted chastisement on him in the presence of the townsmen.'[5]

The farmer finally admitted to the theft after three gods accused him.[6] In addition to the chastisement of the gods, he was given 100 blows with a palm rib by Penherwere, an inspector of the house of the carrying chair. He also had to swear an oath that he would not steal anything else under fear of being thrown to a crocodile.

One interesting aspect that is raised by this record, in particular, is that it is clear that no oracle's decision was fallible, meaning it was acceptable to ask as many oracles as you need to get the answer you want. In the case of Pethauemdiamun, the answers were undeniable, so he had little choice but to accept them.[7]

Another example of asking oracles multiple times is recorded in Letter 14 of the late-Ramesside letters from Deir el Medina, where the residents were worried about their senior scribe Tjaroy, who had accompanied general Piankh on a military expedition into Nubia. At every public procession of the deified Amenhotep, the petitioner asked for the safe return of Tjaroy to the village. The oracle assured him: 'I will protect him; I will bring him back safely, and he will fill his eye with my forecourt.'[8]

It is clear that in processional oracles amidst all the noise and chaos of the public, the god was still able to 'speak' and to be 'heard'. However, the god communicated through movement as it would not be practical for the god to make verbal pronouncements as their voice would be drowned out by the noise. Therefore it would be more precise, at least for processional oracles, for the gods to communicate their answers through priests.

The processional oracle continued in use well into the Roman period, although other more reliable methods of communicating with the gods was preferred. In the Roman period, especially in the Delta region, the processions are identified as the 'appearance of [god's name]', or '[god's name] is the speaking appearance'. In the temple of Phepheros, at the Fayoum town of Theadelphia, dated to the third century, there is a depiction of an oracular procession of Sobek showing a mummified crocodile on a long bier wearing an

Atef crown. The bier is carried by priests and others are facing it, presumably to identify the god's movements in order to answer the questions posed to it.

Although there were a number of ticket oracles in the Fayoum (discussed below), this image shows that the processional oracle still had a place at this late date.[9] The records of processional oracles become rarer after the third century, but not necessarily because they were in decline, but because spontaneous questioning in a crowd did not lend itself well to record keeping, and it was also difficult for the temples to make an income from it.

## Correspondence Oracles

As processional oracles did not happen in every town every week, there could be pressing issues which needed to be dealt with quickly and therefore another method of approaching the gods was required.

One petitioner in the late Ramesside Period (c. 1070 BCE) wanted a personal audience with a god and wrote a note to the temple asking if the god would be available to answer his questions:

> to tell you some matters of mine, (but) you happened to be hidden in your sanctuary and there was no one admitted to send it to you. Now, when I was waiting, I found Hori, this scribe of Medinet Habu, and he told me, 'I am admitted.' So I am sending him to you. Look, you will cast off mystery today and come out in the course of a procession, so that you may judge the matters of the five kilts of the temple of Horemheb and these further two kilts of the Scribe of the Necropolis.[10]

Whilst this pilgrim was hoping the god would make a special procession for him, and even asks someone with access to the god to approach him, others would simply ask their question via written text. From Deir el Medina there are numerous ostraca (pottery sherds) with questions for the oracle written on them. Gods were able to answer yes or no by indicating towards another ostracon with the word 'Yes' or 'No' written on it[11] – rather like a modern Ouija board.

The questions were often rather mundane and refer to day-to-day concerns and were unlikely to be written by people travelling a great distance to the oracle on pilgrimage:

- Is this veal good enough to eat?
- Will the vizier give us a new foreman?
- My Good Lord, is one of my goats with Ptahmose?
- Is it he who has stolen the mat?

Such questions do, however, indicate the importance of religion in the everyday lives of the Egyptians where the gods were treated the same way as you would perhaps approach a trusted elder to get the information required.

By the Late Period (525–332 BCE), this idea of writing to the oracle developed into the rather efficient ticket system where during a procession, two tickets were written with two different outcomes and presented to the god. This system became particularly popular in the Roman period.[12] An example of the types of ticket items are found in a text from the time of the High Priest Pinudjem (108–945 BCE) which provided two statements: 'It is said that Thutmosis, this intendent of goods, is in possession of nothing that can no longer be found.' And, 'It is said that Thutmosis, this intendent of goods, is in possession of something that can no longer be found.'[13] The god is therefore given a choice. In this incident he chose the first example, and then as no oracles were considered fallible, the question was repeated just to make sure.

The ticket system was not a replacement for the processional oracle and an edict from the end of the second century CE states: 'in writings as it were divinely delivered or through the procession of images', showing they were just different options ensuring pilgrims and devotees were always able to get their questions answered.[14] It is thought that the popularity of the ticket system during the Roman era could have been a response to the pilgrims wanting something concrete to take home with them bearing the words of the god. During the pharaonic period, sometimes this idea of a souvenir was presented in the form of amulets they could wear. The ticket system could therefore be viewed as an extension of that.

Oracles were often called upon to provide protection for children. This protection was offered in the form of decrees written on papyrus, rolled up and placed inside hollow cylindrical pendants of gold, wool or leather. They were then worn around the neck. The earliest dates to the Middle Kingdom (2040–1782 BCE), although the early ones have been found empty.[15] Some of these decrees are very simple, such as the twenty-second dynasty (945–715 BCE) example in the Fitzwilliam Museum, Cambridge (E.12.1940) which simply says: 'Words recited by Khonsu in Thebes Neferhotep: he made a goodly protection [for] S3k, the justified.'[16] Despite this simplicity, the papyrus itself was stored in an elaborate gold cylindrical case.

However, a more elaborate oracular amuletic decree, dated to the twenty-first or twenty-second dynasty discovered in one of these amulet cases, is recorded on Papyrus Cleveland 14.723:

Said Khonsu-in-Thebes Neferhotep, the great god, the oldest who was the first to come into existence: I shall protect Irenkhonsu, [son of]

Diuesenmut [*probably mother*], my servant. I shall keep him healthy in his flesh and in his bones. I shall make his dreams good; I shall make those dreams which another man or another woman shall see for him good. I shall protect him from every slander and every injustice thereof. I shall protect him from the demons and from the gremlins. I shall protect him from any who seize a man through capture. I shall protect him from those who seize someone stealthily. I shall save him from a crocodile, a serpent, a scorpion, and from any mouth which bites. I shall protect him from the gods and goddesses on the (*book*) 'That-which-is-in-the-year'. I shall protect him from all the books of the end of the year. I shall protect him from every disorder, sickness, from fever, illness and from flatulence. I shall propitiate for him Montu-Re, Lord of Armant (and) every god and every goddess. I shall guard him from them. I shall provide for him everything good and every good recitation. I shall ward off the bad [oracle] from him. I shall cause that this oracle protect (him) and likewise the oracle which I shall place in the presence [daily].

So, one said namely Khonsu-in-Thebes Neferhotep, the great god, the oldest who was the first to come into existence.[17]

One of the most common oracle souvenirs from the pharaonic period onwards, throughout the temples of Egypt, were gouges taken out of the wall (FIG 17). These date from the New Kingdom through to the fifth century CE with even some more recent examples. These gouges tend to be located on the corners of buildings, hypostyle pillars or specific hieroglyphic inscriptions. Medicinal drinks and salves were created with the dust. The dust was believed to offer the devotee protection if put onto their walls, or fertility if drank or sown into the ground.[18]

## Talking Oracle

Other pilgrims made a journey to the temple to address the god or revered ancestors directly. The talking oracle was suitable for those with questions which had complex answers or for those who were visiting from other towns. As the temples in general were closed to the public, specific areas were dedicated to the talking oracles.

Such a reception area was created on the upper level at Deir el Bahri, the temple of Hatshepsut, on the west bank at Luxor. The temple of Hatshepsut itself had stood in ruins until 260 BCE but these were slowly cleared, with one of the rooms ready to receive pilgrims by approximately 116 BCE. Initially the oracle was dedicated to Amenhotep son of Hapu, who was very popular with the visitors.[19] During the Hellenistic and pre-Roman period this site was then very closely associated with Imouthes-Asclepius, the god of healing and one of his daughters, Hygeia, a personification of health and cleanliness.

The oracle at Deir el Bahri comprised two rooms, the outer room for the visiting devotees and the inner room for the deity. There was a grill above the door between the two rooms, through which the priest spoke the will of the gods to the devotees. Alternatively the deities spoke within a dream, and the priest took on the role of interpreter. This was known as incubation.[20] This incubation oracle was dedicated to the New Kingdom cult of Amenhotep I and remained in operation until the end of the second century CE, and was well known as a place of incubation and healing.[21]

The priests encouraged the pilgrims to leave inscriptions on the walls, as graffiti was not considered the act of vandalism it is today. It was considered a means of broadcasting the success of the oracle and dream therapy to the world and honoured the god.[22] The ancient equivalent of tagging your favourite business on social media.

The records left at the site indicate that the majority of the Graeco-Roman visitors were from the local area; from the Memnonite and Hemonthite nomes and came to the site for religious purposes,[23] rather than as disinterested tourists. They came from all social classes including a donkey-driver and olive-presser which is unusual as they would not be expected to be part of the literate society and therefore generally unable to leave their name on the wall.[24]

Many people at this time also visited the now abandoned Deir el Medina for cures to illnesses or for help with fertility issues, hoping the deified Amenhotep I would be able to help them. Amenhotep I was the king who founded the village and had been deified in the New Kingdom.

There is little doubt that many visitors to these oracles believed the process to be genuine and that they were hearing the words (and therefore the voice of the god).[25] The mechanics of the oracle, whether spoken, processional or written communication was so well carried out that it was difficult, if not impossible, to work out how it was done. However, there were always those who questioned the mechanics and one pilgrim, Athenodorus, left a detailed graffito at Deir el Bahri in Greek outlining how he heard the voice and wanted to know where it was coming from. He opened the door from the pilgrim's room to have a look, hoping to catch the priests in the act, but they were very good at hiding themselves and he was left cured of his illness and a firm believer in the voice of the god.[26]

As time progressed the temples became a little more sophisticated than a 'secret' room from which a priest spoke, and a fourth century inscription from Alexandria describes how:

[The priests] had constructed statues of bronze and wood hollow within and fastened the backs of them to the temple walls, leaving in these walls certain visible openings.

Then coming up from their secret chambers they got inside the statues, and through them gave any order they liked and the hearers, tricked and cheated, obeyed.[27]

Archaeological evidence of these priest holes have survived including examples at the two oracle temples at Karanis in the Fayoum. At the rear of the temples there are large plinths which held mummified crocodiles representing the deity Petesouchos, a local form of Sobek. Beneath these plinths were small rooms, large enough to conceal a priest. These were only accessible from the far side where they could not be seen by the pilgrims addressing the god.

Other examples in the Cairo Museum include four naoi or stone boxes which housed a cult statue of the type which was carried in procession. All of these have large holes drilled into the side which are thought to be a sign of an oracular function.[28]

However, only one statue seems to have been discovered which could have been used for an oracular function in this way. That is the 51 cm bust of Re-Harmachis in the Cairo Museum (Cairo 66143) dated from the Late Period (525–332 BCE). This statue is hollow and has a mouth hole which was placed against the wall with a priest hiding on the other side, speaking through the statue.[29]

What is quite intriguing is that, despite this archaeological evidence, in the New Kingdom it was a specific crime to 'falsify the oracle' as the *Instruction of Amen-em-ipet* stated:

> Do not falsify the oracle on papyrus,
> And so harm the plans of god.
> Do not invent for yourself the power of god
> As though there was no fate and destiny.[30]

What is meant by falsifying the oracle, if it was taken as fact that the gods were speaking through the priests? Unless of course, this did not apply to the oracle priests themselves. The Indictment Papyrus from the reign of Ramses V (1145–1141 BCE) however, discusses charges against two priests of the oracle at the temple of Khnum at Elephantine, who were charged with trying to 'manipulate the oracle', which does suggest that the priests could be guilty of doing this too. How this was judged, however, is not clear.

## Incubation

Incubation was an extension of the talking oracle, although the petitioner was generally induced into a deep sleep, in order to enter the 'next world, the realm of the gods'.[31] Their dreams were then interpreted by the priests at the temple, as being messages from the gods. The earliest evidence of dream incubation dates to the First Intermediate Period (2181–2040 BCE).[32]

Whilst many people went to the incubation oracle to be cured of an illness or to facilitate fertility these were not the only reasons. One Roman, Serenus, received instruction from Apollo in a dream to visit the temple of Philae meaning he then embarked on the long journey from Alexandria to Aswan.[33]

One of the most famous sanatoria was the Ptolemaic temple dedicated to the goddess Hathor at Dendera. This was a mudbrick building to the south of the mammisi of Nectanebo. This early Roman sanatorium comprised private chambers surrounding the central rooms, each with niches for divine statues. Along with water rites and other healing rituals this suggests that healing also occurred through incubation.

## DIY Oracles

Not everyone was able to attend the oracles located at the temples, but that did not mean they did not need divine help. This led to diviners making themselves available within the villages who were able to answer questions according to a book, known as the *Sortes Astrampsychi*. This third century document is fascinating and records between 90–110 typical questions asked by Graeco-Roman petitioners, to each there were ten possible answers. Questions included 'should I become a fugitive' (question no. 86) or 'should one be sold' (question no. 74).[34]

The answers to the questions were spread across a table, and there was a method for the interpreter, or diviner to choose the true answer. This method was to:

1) Choose a random number between 1 and 10
2) Add this to the number assigned to the particular question in the list
3) Match the total to a second number in a table in the book

The interpreter of the divine message did not require much training to provide the right answer using this book, and like many things, conviction may have been in a confident delivery.

It is likely the *Sortes Astrampsychi* was a widely and long used tool in the oracle industry, as at least nine copies have been discovered dating from between the third and the fifth centuries CE. Mary Beard comments:

> Oracular books in the hands of private diviners could circulate widely, reaching where the spoken word of god was rarely or never heard. Any town could have its own alphabetic oracle, ready for the casual use of its citizens on whatever trivial, or not so trivial, matters were troubling them.[35]

## Who Visited the Oracle?

The oracle was for anyone, from the very poor through to the kings and queens of Egypt, as well as for international travellers. The very public elements of the oracular processions meant the gods were available to everyone, but the wealthier you were, the more private the audience you could have with the deity of your choice.

The oracle provided a way for the ordinary people of Egypt (and further afield) to converse directly with a god without having to go via an intermediary whether that be the king or a priest, or make a payment. Of course, if people could pay, they did, and the wealthier they were the more elaborate their payments were. There are examples of inscriptions where petitioners bargained with the gods, telling them what they would provide if their wish or issue were to be resolved. An example from New Kingdom Deir el Medina states: 'Said by Hor-Nefer to his god Amen-Re, Lord of the Thrones of the Two Lands. If I see that you bring it about that I get it, I will make a jar of imported date-beer for you, and also a jar of beer, and so will my man, [and also] … loaves and white bread.'[36] Bargaining and questioning answers shows the very real relationships people had with their gods. Yes, they were divine, but their very human natures were embraced by the population.

For the poor, attending the oracle may have been a special occasion, or even a last resort unless in procession. However, if you were wealthy enough to visit the temple directly, there was no limit to how many times you could address the gods, even about the same thing, as Papyrus Merton (II 81, 3–7) shows. It records that a Roman mother who was concerned about her son taking a journey to Rome, 'consults the oracle every ten days' only to discover, 'that good opportunities were not available [to him] this present quarter'.[37] She was clearly accessing the oracle within the temple rather than on procession and it seems that many oracles changed the in-temple process to include a private audience from the New Kingdom onwards, continuing until the Roman period.

Other people, like the architect in the mansion of Ramses II, contacted the oracle, not to ask for something but as a vote of thanks. This architect had returned to his hometown of Koptos safely after having travelled abroad and thanked Isis for her help in his career as the Chief of Foreign Countries in the north.[38]

The oracle was also used by officials as an integral part of the legal system, and in many cases, whatever a god decreed was taken by the local villagers as legally binding. It was used as a means of solving disputes, but also to guide people through crises and trouble.[39] The written evidence suggests that when it came to legal appeals, quite often the accuser already knew who the guilty party was but did not have enough evidence to convict them. In other cases, the oracle was approached when the accused was wealthier than the accuser, which would have probably gone against the accuser if they applied to the court instead.[40] Often when visitors approached the oracle they asked, 'did so-and-so steal the item?' or they provided a list of names of potential guilty persons, with the oracle indicating who the guilty party was when the names were read out. However, if the guilty party was not already known, it did not prevent the petitioner from asking the oracle for guidance.

At Deir el Medina, the villagers worked as magistrates on a rota system in order to oversee their local law court known as the knbt. This was separate from the oracle, but both were considered equal regarding settling legal disputes. On two occasions the oracle coincided with the holding of the knbt, although whether this was deliberately coordinated is unknown.[41] As with any other oracle, at Deir el Medina, the statue of a god was addressed by the petitioner and whichever way the statue moved indicated the verdict on that particular case.[42] The only real difference is that the decision for legal pursuits were often recorded in full detail including a list of witnesses and the evidence presented. This may have been a means of ensuring the verdict or punishment was carried out, as well as proving the matter had been settled.[43]

One such case is recorded on an ostracon in the British Museum and dates to the fourth year of Ramses IV (1147 BCE) where an appeal was made to the oracle of Amenhotep I, by a person named Kenna, whilst the god was on procession in the necropolis. Kenna was annoyed that even though he had renovated a derelict chapel, an individual named Merysekhmet claimed joint ownership. Merysekhmet had also addressed the oracle previously and was told he could have joint ownership with Kenna, even though the chapel had nothing to do with him. Kenna swore that should his testimony be untrue then he would be liable for 100 blows.[44] The deified Amenhotep I made it clear that the chapel was owned by Kenna and that no one was to have shared ownership.[45]

It has been suggested that in the Ramesside period (1151–1070 BCE) at Deir el Medina, the use of the oracle for clearing up legal disputes was on the rise. Two reasons are given for this: one, the knbt, or court, was becoming increasingly corrupt and the petitioners did not feel they were getting a fair hearing especially if they were not wealthy, and two, there was an increase in personal piety.[46] However, in the Egyptian legal system, unless there was a crime against the state, the knbt had the same legal authority as the oracle. There was no real power to enforce a sentence,[47] and petitioners were able to choose whether they took their legal complaint to the court or to the god. Both were carried out publicly, and often it was left to social pressure to make the litigants comply with judgements, such as returning stolen goods or paying a fine. If they did not comply and the crowd were not that interested, then there was little the court, or the oracle could do. Unfortunately, in the majority of oracular appeals, we know the verdict, but whether these were complied with is very rarely recorded – it is only if records show that they approached another oracle with the same issue that we know they did not follow the original verdict, with the issue remaining unresolved.

## Royalty and the Oracle

It was not only the general populous who approached the oracle for help. There are numerous records to show that royalty visited the oracle, often to demonstrate their legitimacy to rule especially if there was some doubt over their succession. Hatshepsut (1498–1483 BCE), for example, who wanted to rule as pharaoh rather than the female role of queen, went to the oracle of Amun at Luxor Temple, the 'very great oracle in the presence of this good god, proclaiming for me the kingship of the Two Lands, Upper and Lower Egypt … Being the ordination of the Two Lands for me in the South Hall of Luxor Temple while His Majesty [Amun-Re] delivered an oracle in the presence of this good god.[48]

Thutmosis IV (1419–1386 BCE) recorded his oracular experience on the Dream Stela, which he erected at the foot of the Sphinx at Giza. As a prince, while hunting gazelles in the desert, he rested between the paws of the Sphinx and in his dreams, he was told by the god Hor-em-akhet that if he cleared the sand from the paws of the Sphinx, he would be king (FIG 18). Obviously Thutmosis removed the sand as requested, carried out repairs to the body and paws of the Sphinx and repainted the statue in glorious colours to honour the god. Hor-em-akhet was clearly pleased as Thutmosis, indeed, became king.

However, gods did not always side with kings, and Herodotus records that the oracle shrine of Leto (Wadjet) in the Delta city of Buto gave warning of imminent death to King Menkaure (2532–2504 BCE):

> After the death of his daughter, a second calamity fell upon Mycerinus [*Menkaure*]. He received an oracle from Buto to the effect that he was destined to live only for six more years and to die within the seventh. He sent back an angry message to the shrine and reproached the god with the injustice of allowing a man so pious as himself to die so soon, when his father and uncle, who had closed the temples, forgotten the gods, and afflicted their fellow men, had lived to a good old age. In answer to this there was another message from the oracle, which declared that his life was being shortened precisely because he had not done what he ought to have done: for it was fated that Egypt suffers for one – a thing which his two predecessors had understood very well. Mycerinus, convinced by this that his doom was sealed had innumerable lamps made, by the light of which he set himself every evening to drink and be merry, and ceased day or night from the pursuit of pleasure … his object in this was by turning night into day to extend the six remaining years of this life to twelve, and so to convict the oracle of falsehood.[49]

Herodotus also records that the oracle at Buto prophesised to Psammeticus I (664–610 BCE) that vengeance and 'men of bronze' would come from the sea. Psammeticus had apparently been exiled at the hand of eleven kings:

> In bitter resentment at the treatment he had received, Psammeticus planned revenge. He sent for advice to the oracle at Buto (the most veracious in Egypt) and was told in reply that vengeance would come from the sea, whence bronze men would appear. Psammeticus did not believe a word of this and thought it most improbable that he would get any help from the bronze-men; but not long afterwards it so happened that a company of sea raiders from Ionia and Caria were forced by bad weather to land on the Egyptian coast.[50]

Another inscription (JE72130) from Hermopolis dated to the reign of Nectanebo I (380–362 BCE), refers to the king when he was a general and approached the oracle of Nehmatawy at Hermopolis. The oracle proclaimed 'that he would be King of Upper and Lower Egypt for many years as a good ruler of this land'.[51] Nectanebo then went back to the royal residence and told Achoris/Nepherites II (393–380 BCE) of the prediction, which subsequently resulted in him being crowned king. It is likely that he already had a claim to the throne as the grandson of Nepherites I (399–393 BCE) and this oracular prediction was extra evidence supporting his claim to the throne.[52]

However, a king did not just approach the oracle about his own succession, and there is evidence that Ramses II (1279–1212 BCE) approached the oracle of Amun-Re to ask about the next appointment of a new high priest of Amun. The god indicated that Nebwenenef would be the ideal candidate.[53] The confirmation of the role of high priest of Amun was made by the oracle of Amun at least until the twenty-second dynasty.

## Gods of the Oracle

When considering the question of which gods were popular for the oracle, the answer could be 'any of them'. Wherever a god was revered locally, and people came to the temple or to the god for help, an oracle could be created by the priesthood of that deity. It was a supply and demand service. However, that is not to say that there were not famous oracles, to which people travelled from across Egypt and in some cases across the ancient world to address. There were national deities like Amun, and international deities like Isis for example who were a major draw for pilgrims.

### *Isis*

Isis was a popular oracular goddess, particularly in the Roman period as she was actively worshipped within the Roman pantheon. Therefore, there were many oracles dedicated to her which survived longer than those of other deities. Diodorus Siculus, writing in the first century CE describes Isis' power:

> Isis … finds her greatest delight in healing of mankind and give aid in their sleep to those who call upon her, plainly manifesting both her very presence and her beneficence towards men who ask her help … For standing above the sick in their sleep she gives them aid for their diseases and works remarkable cures upon such as submit themselves to her. [History 1.25.3.5]

This suggests the Isis oracles were primarily incubation oracles, where dreams were sent to pilgrims who slept within her temples. One such incubation temple was at Menouthis, on the north coast of Egypt, dating to the third century CE and surviving until 414 CE.

The version of Isis worshipped here is likely to be in her Hellenic form rather than the traditional pharaonic form, and she is referred to as 'she who hears prayers'. She is shown with an Atef crown, normally associated with Osiris, and a pair of hearing ears. One Alexandrian philosopher who stayed at the temple records that he received:

> an oracle (or rather he was deceived by the demon appearing as Isis) according to which the goddess promised him children if he went with his wife into the temple that the goddess had in Menouthis, a village 40 miles from Alexandria.
>
> He stayed some time in Menouthis and offered a considerable number of sacrifices to the demons. But it was to no avail. The sterility of his wife persisted, nonetheless. Having believed that he saw in a dream Isis lying beside him he heard it declared by those who interpreted dreams there and who served the demon expressed in Isis, that he ought to join himself to the idol of that goddess, then to have sex with his wife – that thus a child would be born unto him.[54]

When this did not work, he was sent by a priest to the nearby town of Astu where a local priestess would give him a baby. It is interesting that whilst on one hand he refers to Isis as a 'demon', on the other he was willing to try anything to have a child showing there was a modicum of belief in the system, if not the deity.

Another temple of Isis was at Philae which was closed in 540 CE. Even though the religion in the region of Philae had been overtaken by Christianity, eschewing the pagan gods, Christian pilgrims still visited the site to ask Isis for help. Many also visited the temple as tourists, when they were on pilgrimage on the nearby island of Elephantine (see chapter 4).[55] The temple of Isis was maintained and staffed by the Nubian Blemmyes who ensured its endurance into the sixth century.[56] Eastern Roman diplomat and Greek historian and rhetorician Priscus (fifth century) describes a ritual at Philae where the Blemmyes:

> were to have the right to cross unhindered to the temple of Isis, while the Egyptians had the care of the river boat in which the statue of the goddess was placed and ferried across the river.
>
> At a fixed time the barbarians take the statue across to their own land and, when they have taken oracles from it, return it safely to the island.[57]

However, even Isis who was one of oldest Egyptian deities was not immune to changing popularity when it came to religion.

There was a cult temple dedicated to Isis at Canopus in the Delta, which had been founded in the Ptolemaic period. Initially, her cult was in direct competition with the incubation cult of Serapis, also at Canopus. Serapis was a Graeco-Roman deity who was introduced by Ptolemy I (305–282 BCE) and rose in popularity. However, the Isis cult eventually eclipsed the cult of Serapis as she became more popular despite the importance of the Serapeum in Alexandria.[58] Strabo (63 BCE–23 CE) described the importance of the Serapis cult:

> The temple of Serapis, which is honoured with great reverence and effects such cures that even the most reputable men believe in it and sleep in it – themselves on their own behalf or others for them. Some writers go on to record the cures and others the virtues, of the oracles there. [17.1.17]

When the Isis temple at Canopus temple was closed in 389 CE the priests simply relocated to the Menouthis temple, where the cult of Isis continued to gain in popularity. As Christianity grew in the region, the incubation cult of Isis morphed into the dream cult of saints Cyrus and John (chapter 4) which attracted a different kind of pilgrim, whilst continuing a long pagan tradition.[59]

## *Apis Bull*

The Apis bull cult was centred in the Serapeum in Memphis and was revered as an important oracle. As well as attracting local pilgrims, there were several visitors from further afield who wanted to see the spectacle of the Apis bull. The Apis bull was an important deity in the Egyptian pantheon who had been worshipped since the first dynasty (3050–2890 BCE) and was associated with the goddess Hathor. The bull was identified by a white triangle on its forehead and black patches resembling birds on his hide.

The sacred bull was allowed out of his pen once a day to frolic in the fields, although his movements were carefully watched and recorded. The priests interpreted his movements to the pilgrims who gathered to witness it.[60] Strabo records:

> Into this court they set Apis loose at a certain hour, particularly that he may be shown to foreigners; for although people can see him through the window in the sanctuary, they wish to see him outside also; but when he has finished a short bout of skipping in the court they take him back again to his familiar stall.[61]

Also at the Serapeum there was a temple enclosure for the mothers of the Apis bulls and sacred falcons. In a small brick side chapel, it is thought a statue of a baboon, associated with the god Thoth, the god of wisdom and writing was placed and also questioned by visitors as an oracle.[62]

The gods Khonsu the Contriver and Khonsu-who-was-a-child are mentioned in six oracular decrees and are both in baboon form, seated alongside another oracular deity, Khonsu-in-Thebes Neferhotep. It is thought this form of oracle may have been located in Karnak in a shrine where the statues of all three forms of Khonsu were represented.[63]

## *Buchis Bull*

A similar sacred animal to the Apis bull was the Buchis bull at Armant which was first mentioned in the thirtieth dynasty (380–343 BCE). The Bucheum Stepa 9 (JE53147) records that the Buchis bull was chosen primarily according to its colouring with the priests from Armant travelling to a potential new Buchis bull to inspect him, wherever it may have been within the country.

To aid with the inspection, there was a procession with the sacred barque of Amenope of Djeme, which only stopped and was immoveable when it was in front of the chosen bull. Amenope of Djeme was the main god of Luxor Temple where the Buchis bull would eventually be enthroned and was likely an incarnation of the deified Amenhotep I.

Although the priesthood could recognise the markings, only the power of the god himself could truly recognise the sacred bull.[64] Once correctly identified, the Buchis bull was transported in a glorious procession to Thebes, with the king travelling from Alexandria in order to witness the spectacle, and there may have been a public element of the procession which would have been a draw for localised pilgrims.

There are other records of the sacred barque stopping dead as a means of identifying a specific location and it seemed to be a legitimate way of identifying ideal tomb or temple locations. For example, Thutmosis III records that when he was a child, attending a procession at the hypostyle hall at Karnak, when the sacred barque of Amun stopped in front of him and publicly recognised him as the next king.[65] Unfortunately, within the temples such processions would be inaccessible to the general public.

## *Sobek*

The god Sobek, the crocodile deity normally associated with Kom Ombo in the south of Egypt became a national deity from the Late Period (525–332 BCE) until the fourth century CE. For example, there were at least three

Sobek oracle temples in the Fayoum which generally followed the ticket system, and a number of tickets have been discovered written in Demotic, the Egyptian language in Greek letters. This suggests that many people visiting the temples were Egyptians.

One Sobek temple at Tebtunis in the Fayoum has surviving evidence of a priesthood functioning well into the fourth century showing the oracle was likely still in operation.[66] Many of the papyri decrees surviving from the oracle look like legal contracts. The petitioner declared they would work at the temple as a servant, as well as paying a monthly fee to be protected from supernatural forces: 'protect me, guard me, protect me from every evil spirit, every sleeping man, every drowned man, and every dead man'.[67] It appears however, that all the petitioners at Tebtunis were children, whose father was unknown, and this oracular decree could have been a way of making illegitimate children wards of the temple under the protection of the gods.

The cult of Pnepheros and Petesuchos twin aspects of Sobek at Karanis, in the Fayoum also had a voice oracle from the beginning of the temple's beginning under Ptolemy II (285–246 BCE) to the third century CE.[68]

## *Mandulis*

At the cult centre at Talmis (Kalabsha), in the south of Egypt, there was an incubation oracle until the third century CE dedicated to a local Nubian god, Mandulis. Due to his close association with the Greek god of time, Aion, it attracted several Greek and Roman visitors to the site. This international connection spread the reach of what would otherwise have been a very localised cult.[69]

It was initially built in conjunction with a Roman garrison as a common holy site for Roman soldiers and was in use from the beginning of the Roman period (30 BCE) to the fifth century. As well as being associated with Aion, Mandulis was also called the son of Horus which further increased his popularity with Roman pilgrims. One such pilgrim left an inscription:

> Oh, ray shooting lord Mandulis, Titan, Makareus, having beheld some radiant signs of thy power I pondered on them and was busied therewith, within to know with confidence whether thou art the sun god …
>
> I had a vision and found rest for my soul. For thou didst grant my prayer and show me thyself going through the heavenly vault; Then I knew thee, Mandulis, to be the Sun, the all-seeing master, king of all, all powerful Eternity [Aion].
>
> O happy folk that dwell in the city believed by the sun Mandulis.[70]

Another inscription left by one Herodas asks the oracle for a safe homecoming while a soldier named Maximus wanted success in the army.[71]

The incubation cult at Talmis seemed to form only a short part of its history, with inscriptions attesting to this function only for the first three centuries. Then the graffiti changes showing that pilgrims were now primarily young men, some from Greece and Rome, in search of heavenly knowledge as a means of learning about the gods' true essence. They had moved away from healing to self-development and spirituality.[72]

As part of the 'touristic' element of this temple, with their pilgrims travelling long distances to be there, the priesthood took the opportunity to create small terracotta statues of the god which they could sell, giving the pilgrims something to take away with them as a memento of their trip.[73]

## Christian Oracles

When Christianity reached Egypt in the first century CE, there was a juxtaposition for some time of traditional religious practices and the newer Christian ones, and this meant that oracles became an important aspect of Christianity in Egypt. For example, many pilgrims travelled to the oracle of Saint Cyrus and Saint John which was situated to the east of Alexandria at Canopus, replacing the older cult of Isis.[74] Although the religion was different, the questions asked of the Christian god did not vary much, including such things as asking for a child from the oracle of Saint Leonitus (Papyrus Rylands 100), and advice on a betrothal (Berlin 21269).[75]

The Christian oracles generally followed the ticket system where two conflicting sentences were put to the god:

> My Lord God Almighty and Saint Philoxenus my patron, I beseech you through the great name of the Lord God, if it is your will and you help me to get the banking business, I invoke you to direct me to find out and to speak.
>
> My Lord God Almighty and Saint Philoxenus my patron, I beseech you through the great name of the Lord God, if it is not your will for me to speak about the bank or about the weighing office, direct me to find out that I may not speak.[76]

As discussed in chapter 4 in more detail, incubation was also an important aspect of Christian pilgrimages. The closer to the relics the pilgrim was able to sleep, the more effective the cure would be for ailments, or the advice received from martyrs (not God directly).

## Oracle as a Business

With such a long history there is little doubt that the oracle started as a religious practice allowing the general populace to communicate with their deity and gain the advice and protection they needed. However, by the Roman period, the oracle was big business, and provided a lucrative income for the temple.

Processional oracles whilst still having a place in the community were too chaotic and random a system to attract pilgrims from afar. Additionally, processions were more ad hoc, they could be expensive to put on and were time consuming. Therefore, many were replaced by a stationary oracle, often appearing in areas with no previous tradition of an oracle. Stationary oracles were easier to manage and could potentially attract more people, making it easier to monetise them.

This approach also saw the introduction of new oracle cults, like the oracle of Bes at Abydos (chapter 8). Whilst being complementary to local traditions and cults his international popularity guaranteed pilgrims would travel to approach him.

During the Graeco-Roman period pilgrims were often young Romans and intellectual pagans[77] and they wanted to have something to show for their trips. This led to the ticket and the correspondence system which meant the pilgrim could take away a note from the god which was an upgrade of the earlier amulets, incubation and healing practices.[78] With the changing visitor types, the questions asked of the gods also varied from the mundane day-to-day questions of the local population to the questions of the well-heeled young Roman who travelled to the oracle as a tourist/pilgrim.[79]

During the Roman period (and indeed earlier), local and regional shrines and temples were reliant for the most part on gifts from patrons for their survival, due to a declining imperial patronage. It was, therefore, beneficial to have an oracle which appealed not only to the local community but also to well-off Roman pilgrims and politicians. Lucian, a Roman satirist, mentions 'the price set for each oracle' which shows there was a cost to address the god, in addition to any donations or construction work the god advised the pilgrim to carry out.[80]

There has been some discussion about which language the oracles answered in, as with many of the visitors being Greek or Roman, they were unlikely to understand the Demotic language commonly used for oracles until the third century CE. However, as oracles were presented almost as a 'cultural package' with priests and the gods behaving in a particular way and the requests mostly written in Egyptian, it is unlikely that a visitor would demand a change in the language of their response to take away as it could reduce the power and authenticity of the experience.[81] It also has to be considered that many visitors to the oracle may have been illiterate in Egyptian and Greek and therefore it would not matter much to them what language the response was written in.[82] The priest was there to interpret the god's words, and the ticket to take away was like a souvenir.[83]

As with any business, there were a lot of oracle cults in the Roman period, all marketing to the same pilgrims. This saw an element of competition between traditional local cults and the Graeco-Roman regional or international cults in regard to power and authority. As with modern tourism, it is the well-marketed sites which get the visitors, those which cater to their needs and create a pleasant experience. In the Graeco-Roman period it was the same, travellers went to the bigger shrines and oracles rather than small local shrines. For example, one person travelled to the oracle of Kysis (Dush) near the Kharga Oasis to ensure the maternity of their wife rather than head to their local shrine.[84]

There were several key sites which were internationally known, and therefore they were more common stops of eager international pilgrims.

### Siwa Oasis

One of the best-known oracles is that of Jupiter-Amun at Siwa, in western Egypt. Although it gained its reputation in the Graeco-Roman period, there had been an oracle here from the eighteenth dynasty (1570–1293 BCE). However, much of the evidence comes from the height of its popularity and the earliest structure here dates to the reign of Nectanebo I (380–362 BCE).

Nearchos records his trip to Siwa after visiting 'the region of the cataract', in the first or second century CE and describes it as: 'where Amun utters oracles for all mankind; and I sought auspicious words, and I inscribed the names of my loved ones on the temples as an eternal devotion'.[85] The most famous visitor to the site, however, was Alexander the Great (332 BCE). Apparently when he entered the shrine with the statue of the god, Amun recognised him as his son and predicted great success throughout the world. Although Son of Re was a standard epithet of a ruler in Egypt, this acknowledgement was recorded as being an actual deification of Alexander which he then used throughout his life.[86]

By the time Strabo was writing, the Amun oracle at Siwa was 'nearly abandoned',[87] although despite this, it entered a state of legend with many classical poets including it in their work.[88]

## Abydos

As discussed in chapter 8, Abydos was a popular pilgrimage site, particularly in the Middle Kingdom (2040–1782 BCE), but it was also the site of an oracle, popular until 359 CE when it was closed down because the emperor (Constantius II) thought some of the questions posed were anti-Roman.[89] The site of Abydos was generally associated with the god Osiris and this had been the focus of the previous pilgrimages here, but the Memnonium of Sety I was dedicated to seven gods – albeit with Osiris being the most prominent.

Pilgrimage to Abydos increased dramatically in the Ptolemaic and Roman periods as much of the tourism was considered to be pilgrimage. During the Ptolemaic period, many tourists came to visit the temple of Sety I as the site of an oracle cult and as a place of healing.[90] This meant the temple had a much wider appeal than the earlier funerary associations as pilgrims were able to visit the site, embark on a period of incubation and interact with the gods directly.[91]

Many visitors left graffiti at the temple in Greek, Cypriot syllabary, Carian, Aramaic, Phoenician and Demotic showing they came from far and wide.[92] Records show there were at least three visitors from Middle and Upper Egypt, two from Alexandria, one from Pelusium,[93] seventeen from Crete, Cyprus and the Aegean Islands, fifteen from Asia Minor, and nineteen from mainland Greece.[94] Much of the incubation graffiti was on the staircase at the back of the temple and indicates this could have been where pilgrims were asked to sleep. One piece of graffiti left by some Galatians show their minds may not have been entirely on the dream messages of the god, as it recorded 'they caught a fox here' indicating a little bit of hunting was also on the cards.[95]

In the Roman period the voice of the oracle in the Memnonium was the voice of Bes, the dwarf god often associated with childbirth. Bes was associated with sleep from the pharaonic period and protected sleepers and dreams and there are references which, 'Request for a dream-oracle of Bes' indicting he was at the centre of an incubation cult. In Ptolemaic Egypt Bes was said to 'repel the demon-prowlers by night as well as by day'.[96]

Amongst the twenty-eight pieces of graffiti from the fourth century at Abydos attesting to the Bes oracle, Ammianus Marcellinus tells of his experience:

> In the furthest part of the Thebaid there is a town called Abydos, where a god locally called Besa used to reveal the future through an oracle and was worshipped with traditional rites by the inhabitants of the surrounding regions.
>
> Some of those who consulted the oracle did so in person, others sent a letter by an intermediary containing an explicit statement of their requests.

In consequence, records of their petitions on paper or parchment sometimes remained in the temple even after the replies had been given.

Some of these documents were sent to the emperor [Constantius II] out of malice. His small mind made him deaf to other matters, however serious, but on this point he was more sensitive than the proverbial ear lobe, suspicious and petty. [19.12 3–6][97]

There is little evidence of an actual cult shrine of Bes at the site although there was a priesthood which ran the oracle. It is likely they dealt with international correspondence to the god, as well as controlled the sacred space within the temple when pilgrims came to visit.

The location of the graffiti left by pilgrims shows the areas they were permitted to go, with 25 of 29 inscriptions dedicated to Bes inscribed on the southwest exterior wall just outside the Osiris chamber in the Memnonium indicating they were not permitted into the presence of the god per se.[98] This graffiti also tells us that some visitors returned again and again, for example Veristimus who visited five times, signing his name once in the cella of Isis and four times in the second hypostyle hall. It is possible that he was visiting on behalf of others although he does not explicitly state this.[99] Sometimes visitors left their job titles as well as their names, and many of the dedications in the Roman oracle temple dedicated to Bes were athletes from the Olympic and Pythian Games.

The reason the household deity Bes suddenly became an oracle here is unclear as there is no evidence of an older birth chamber in the area which would show a continuity of devotion. It is thought to be an 'example of a domestic cult with transregional recognition that was subsequently elevated to the status of transregional cult'.[100] It could be that the entrepreneurial priesthood here may have founded the oracle of Bes, as a way of boosting visitors to the site by drawing on an established deity. Changing the oracle from Osiris to the more popular Bes in the first and second centuries CE may have been just viewed as a business transaction.

The Memnonium had functioned as a centre for Osiris worship and there were two halls which may have been designed to host a number of pilgrims. It appears that these rooms were accessible to the pilgrims as there is a great deal of Christian damage indicating they may have felt threatened by the popularity.

## Horus-of-the-Camp

In addition to the large national shrines which attracted visitors from all over the world there were several smaller local oracles, which may have had a smaller reach due to the local nature of the deity. For example, in the Third Intermediate Period (1069–525 BCE) Horus-of-the-camp was a small rather obscure oracle, near modern El-Hiba, south of the Fayoum. No other temples to this deity have been identified, although there is a reference to him on the pylon of the temple of Khonsu at Karnak.

There were military installations at El-Hiba, therefore this particular incarnation of Horus was likely related to this military camp. However once the military camp was destroyed in the twenty-first dynasty by the Theban dynasty of high priests, the deity disappeared into obscurity.

The ticket system was used at this oracle where two statements were presented with different outcomes or answers and the god chose which was correct. The pilgrim received the correct statement back.[101] On some examples of petitions from this oracle, found in a tomb in Naga el Deir, one of the petitions is underlined in red,[102] although it is not specified whether this was the accepted or rejected scenario, but it was likely known to the recipient.

Petition A (underlined in red):

Horus-of-the-Camp, my good lord! Concerning this cow of which Teuhrai said: Karsasi son of Hartenu gave it to Pameshem in return for its price, and Pameshem paid him the first instalment of its price and he withheld the remainder. You say: Pameshem is right! He completed the payment of the cow.

Petition B:

Horus-of-the-Camp, my good lord! Concerning this cow of which Teuhrai said: Karsasi son of Hartenu gave it to Pameshem in return for its price, and Pameshem paid its price likewise. You say: Karsasi is right! Pameshem did not pay him the full price in return for his cow.

It has been suggested that when the oracle gave his answer then the priest read out the correct statement, therefore putting the statement firmly into the mouth of the god.[103]

The oracle was clearly an important means of approaching the god and getting the guidance and help required from the pharaonic period through into the Christian period. However, oracles were not the only pilgrimages which were important to the pharaonic population of Egypt. In the next chapter we go back to the pharaonic period to investigate their pilgrimages and other more pedestrian reasons for travel.

# Chapter 8

# Pharaonic Travel

One author wrote in 1916 that 'It's hard to imagine the Egyptian as a tourist.'[1] Whilst in the Graeco-Roman period Egyptians travelled, as we move further back in time, identifying travel and tourism becomes more difficult for two main reasons: people may not have travelled as often or as far, and, due to reduced literacy levels within the community, such trips were simply not recorded. However, through literary sources, religious records and graffiti we can present a rough outline of travel in the pre-Graeco-Roman era.

The literary sources suggest that there was a certain distrust in travelling in the pharaonic period as an activity and a New Kingdom poem from Deir el Medina, states:

> Behold, I do not want to depart from Thebes.
> Save me from what I abhor!
> Every time I leave on a journey,
> I travel north,
> With the city beside me,
> With the Temple of Amun on my path (may I reach it!),
> Medamud before me,
> And Tenet Khonsu together with me
> In the boat of my mission.
> Bring me to your city, Amun!
> Because I love it.[2]

This poem, whilst making it clear that Thebes was the author's hometown and the place they loved best, they do discuss travelling north by boat and some of the places they saw on the journey showing the journey was a familiar one – to the author and the audience. Writing about a desire to stay home does indicate they may have experienced the opposite – travelling away from home.

This idea of preferring home soil is further emphasised in the common greeting 'may you die in your town', which to the modern ear does not sound that cheery, but the idea of dying on foreign soil was abhorrent to the Egyptians. The fact it had slipped into common vernacular indicates that it may have been a common enough event, especially for those in the military.

As much of the literature comes from Thebes, the journeys described are generally northwards and the theme appears often enough in New Kingdom love poetry to suggest it was common, but not always easy:

> I fare north in the ferry
> By the oarsman's stroke,
> On my shoulder a bundle of reeds;
> I am going to Memphis
> To tell Ptah, Lord of Truth,
> Give me my sister tonight.[3]

The narrator in this poem is not in his own boat as he travels northwards as he is being rowed by an oarsman. This indicates boats could be hired for journeys north (and south) as they were in the Graeco-Roman period (see chapter 6) for those who did not own their own vessel.

Other literary texts which indicate that travel was not unknown include the *Shipwrecked Sailor* dating to the Middle Kingdom (2040–1782 BCE). The story recounts the adventures of an official who is about to explain to the king why his expedition had failed. He is nervous and his attendant tells a story starting with, 'Take heart my Lord! We have reached home.' The epitome of good luck: returning to Egypt. He then tells a tale of his expedition where he was shipwrecked upon a marvellous island whose sole inhabitant was a giant snake. The island was abundant in every good thing and when the sailor spoke of bringing incense and offerings to the snake he laughed claiming, 'I am the Lord of Punt, and myrrh is my very own. That *Hknw* oil you spoke of sending, it abounds on this island.' An interesting and convenient aspect of the story is that the snake makes it clear that the island will disappear once the sailor is rescued meaning that no one would ever be able to find it again should they try.

Another Middle Kingdom tale which gives an insight into the notion of pharaonic travel is *The Eloquent Peasant*. The peasant is from the Wadi Natrun in the Delta, and he travelled to Herakleopolis, more than 200 kilometres south, carrying all the 'good products of the Wadi Natrun'.[4]

He took this journey by donkey, laden with the goods, although the text does not tell us how long he travelled for. Part of the journey, just before reaching Herakleopolis took him through the barley field of Nemtinakht, which ran parallel to a canal. Nemtinakht decided to steal the peasant's donkeys and goods and beat the peasant for good measure with a stick. This demonstrates that travelling was a particularly rural endeavour with a lack of official roads and shows the potential dangers. The rest of the text is about the peasant appealing to the officials over this robbery. He made nine separate appeals before he was recompensed for the theft. The officials were more surprised at the eloquence of the peasant than the robbery itself.

What is particularly interesting with these Middle Kingdom stories is the social groups of the travellers. In the *Eloquent Peasant*, the traveller is a peasant, albeit one that owns donkeys, but still a member of the peasant class. This shows that then, as now, if you had the means to move, travelling was open to everyone. Conversely, the shipwrecked sailor was someone travelling as part of his job, and this was a large part of travel in the pharaonic period.

## Travelling for Work

In the modern world we think little of travelling for work, whether nationally or internationally and this has been a trend which has been in place globally for thousands of years. As we saw in chapter 7, officials from Greece, Rome and Nubia were travelling to Egypt on diplomatic trips and, to a certain extent, this was happening in the pharaonic period too. Granted the ancient Egyptians were not jet setting across the world, but there is evidence that not everyone worked in or near their main residence.

In the twelfth dynasty, from the reign of Amenemhat I (1991–1962 BCE), a landowner called Hekanakhte lived in the north of Egypt near Memphis, some 670 kms from his family and home, south of Thebes. At the time, there was a famine in the north and Hekanakhte exaggerated this to his family by saying, 'see they are beginning to eat men here', as a means of showing how good they had it in the south. He emphasised that what he provided them was better than elsewhere, and 'there are none to whom this kind of income is given in any place'. He then went on to explain how his family will have to survive on reduced rations for the time being:[5] 'But as for anyone who will reject the rations among the women or the men, he should come to me here with me and he shall live as I live.'[6]

In a series of letters we learn about the dynamics of his family and the business set up. Hekanakhte was married to Hetepet but planned to bring a second wife into the household. This second wife had not been accepted by the household itself and in one of the letters the maidservant, Senen, is identified as having insulted her. Hekanakhte demanded that she be fired before he called for his second wife to travel to Memphis to be with him.

In another letter to Merisu, Hekanakhte talked about his desire to rent 20 aroura of land (55 square kilometres) in Pr-hA3', a place which has not yet been identified. He intended to plant half with emmer wheat and the other half with barley. As payment for the land rental, Hekanakhte sent some locally woven cloth, for the going rate (24 deben of copper [2.2kg]), as a guide price for Merisu, who appeared to be in charge of Hekanakhte's household and general estate management in his absence.

Hekanakhte already had business dealings in Pr-hA3' as there were a number of people there who had outstanding debts. He told Merisu he had sent two debt collectors, Nakhte and Za-Neb-Nuit, to retrieve payment from debtors in Pr-hA3' and also in another town call Hut-Ha3'.[7] Hekanakhte wanted the debts paid in barley but also presented an amount due if the debt was to be paid in oil instead. It is possible that the debts were due to a poor harvest, meaning the farmers renting the land did not get the yield they needed to pay rent.[8]

The messenger who carried the letters, Za-Hathor, planned to stay in the south for the harvest season and then return to Memphis once it was over. This indicates that it was not that difficult for people to travel backwards and forwards between Thebes and Memphis if they had the means to do so. Despite this ease of movement, Hekanakhte does not seem to have travelled back to the south very often, but, instead, sent the messenger to do the work.

However, these letters never reached the intended recipients, and were found hidden in the tomb of Msh, the vizier of Nebhetre-Mentuhotep, although there was no connection between the letters and the tomb owner. It is thought that the messenger Za-Hathor had been intercepted, probably because of the bulky package of the cloth, which was with the letters, and that thieves took the cloth and threw the letters away, echoing the dangers of the *Eloquent Peasant*. It would be interesting to know what the consequences were as Hekanakhte was very clear in his letter to Merisu: 'Guard all my property! Lo, I will hold you responsible for it! Apply yourself concerning my property!'[9]

An interesting addition to the group of letters from Hekanakhte was the letter from Zat-Neb-Sekhtu who was writing to her mother. She was not related to Hekanakhte and seemed to just take advantage of the messenger going south. The letter was addressed to Gereg, the administrator of her estate, asking that the contents be passed to her mother. The letter included accounts from her estate, indicating that Zat-Neb-Sekhtu was also a landowner, as there was mention of agricultural holdings producing grain and a textile production element. Zat-Neb-Sekhtu also outlined the rations which were to be given to the employees on the land. It is likely that Zat-Neb-Sekhtu was unmarried or a widow as it would have been very unusual for a married woman to live away from the home in this way. It is clear that Hekanakhte and Zat-Neb-Sekhtu chose to work away from their home village, purchasing land in different parts of the country. However, there were a number of jobs which meant some people had little choice about where they travelled, as they went where they were needed.

The *Satire of the Trades* was recorded on a nineteenth dynasty (1298–1185 BCE) papyrus, although snippets have been found from the eighteenth dynasty. It is a fascinating text which outlines the pitfalls of all professions other than that of scribe. This instruction text was written by Dua-khety, to convince his son Pepi that entering the scribal profession was the right choice. Some of the jobs he describes were transient in nature. Take the travelling barber for example:

> The barber barbers till nightfall,
> He betakes himself to town,
> He sets himself up in his corner,
> He moves from street to street,
> Looking for someone to barber.
> He strains his arms to fill his belly,
> Like the bee that eats as it works.

The idea of travelling from town to town to cut hair where it was needed was presented as tiring and hard work. It gives us a brief insight into the life of the barber, where instead of knocking on doors, he set 'up in his corner' and the local people knew where to look for him should they need him. It also hints at a very public activity, and not one of luxury and privacy that salons promote today. There is an image of the barber at work in the Theban tomb (TT56) of Userhet (FIG 19) showing the clients waiting under a tree for the barber to get to them.

Hair, in general, was more important in ancient Egypt than such a rudimentary system would suggest. It could be an indicator of wealth, status and age. Most people, male, female and children, shaved their heads as this was an easy way to combat head lice, meaning the travelling barber in any village no doubt would have been busy. Wigs or linen headdresses were then worn over the bare skin.

Tomb art, and royal mummies, indicate pre-pubescent children had completely shaved heads apart from a lock of hair hanging down the right-hand side. This was shaved off when they reached adulthood. Some New Kingdom priests, similarly, had a plaited side-lock to show their affiliation with a particular deity, such as Khonsu, who is also shown with a side-lock indicating his youth.

However, as children became adults, hair, especially for women, became an erotic element of their appearance. In literature, walking in on a woman dressing her hair or in the process of putting her wig on was similar to walking in on her in a state of undress. The *Tale of Two Brothers* (119–1193 BCE), for example, has the younger brother interrupting his sister-in-law as she was arranging her hair, creating a charged sexual atmosphere, which she totally misread. Papyrus Harris has a love poem which laments:

> My heart thought of my love for you, when half of my hair was braided.
> I came at a run to find you and neglected my hairdo.
> Now if you let me braid my hair, I shall be ready in a moment.[10]

For those who wore wigs, these were maintained within the household or at wigmakers. Such a workshop was discovered in Luxor and soap for washing the wigs made of natron and soda, as well as a substance which could have been used to dye the wigs, was found. Some people shaved their heads and faces at home using copper razors, but others visited the travelling barber when he was in the village. Not only would the job be done quicker and more professionally, but it was an opportunity to pick up gossip from the other villages he visited. Did perhaps the travelling barber take messages and packages to different villages on his rounds?

Another profession that required moving between villages and towns was the courier who may have travelled the full length of the Nile delivering messages, like those of Hekanakhte. Dua-khety tells us that the courier:

> goes into the desert,
> Leaving his goods to his children;
> Fearful of lions and Asiatics,
> He knows himself only when he's in Egypt.
> When he reaches home at night,
> The march has worn him out,
> Be his home of brick or cloth
> His return is joyless.

Not only does Dua-khety emphasis how exhausting travelling is, but he also hints at the dangers of moving outside the safety of the borders of Egypt into the desert. As mentioned above, living (and potentially dying) outside Egypt was considered almost a curse, but so sad is the life of the courier that even the joyous act of returning home cannot lift his spirits.

Other travelling workmen included the reed cutters who were required to travel to the Delta to get the reeds to produce arrows to sell:

> The reed-cutter travels to the Delta to get arrows;
> when he has done more than his arms can do,
> Mosquitos have slain him,
> Gnats have slaughtered him,
> He is quite worn out.

Again, this job is described as exhausting and the Delta marshes as being overrun with mosquitoes and gnats. This description of the location could be Dua-Khety's first-hand experience as he wrote this instruction text as he was travelling south, perhaps from the Delta.

## Soldiers

One of the main ways to see Egypt through work was through the army, but there was no full-time army until the eighteenth dynasty, and the start of the New Kingdom (approx. 1570 BCE). Prior to this, men were rallied if they were needed for military engagement, expeditions or peace keeping. They were trained as the images in the Middle Kingdom tombs at Beni Hasan show, but they were not, at this time, career soldiers.

The army in ancient Egypt is fairly well documented with a combination of evidence regarding their movements ranging from battle reports to graffiti to literary tales. Soldiers travelled nationally and internationally as part of their role and there is graffiti as far south as Elephantine left by Greek mercenaries under the reign of Psammeticus II (595–589 BCE), as part of a campaign to the second cataract and Abu Simbel.[11]

For many soldiers much of their time was spent protecting trade routes, fortresses or the borders of Egypt, and whilst not travelling for pleasure, they were definitely away from home.

For some in the New Kingdom this time was spent in the remote desert mountains high above the Valley of the Kings where several rudimentary huts have been discovered, along with graffiti counting off the days until their 24-day tour of duty ended. During the Amarna period (1350–1334 BCE), in the absence of battle or any military campaigns, the military's key role was to protect the borders of the Western Desert, northern border and the Delta ports where they ensured the safe passage of imported goods into the cities.

Soldiers were also used as protection on trading expeditions where they accompanied the transportation of large blocks of stone for building projects along the Nile, or they dismantled boats across the desert to the Red Sea for trading expeditions to Punt. Additionally the military were also drafted in to help with harvesting in the same way the peasants were drafted to help with state construction projects, such as quarrying minerals from the Eastern Desert and the Sinai. They were able to protect workers from bandits but could also help with the manual labour.[12] One inscription from Gebel Silsila (north of Kom Ombo) outlines the role.

> The first occasion on his Majesty commanding ... in order to execute all the work projects from Elephantine and Samnehdet (in the Delta). The commanders of the expeditionary force were set to carrying out a large compulsory labour project, consisting of quarrying sandstone in order to make the great benben-stone for Horakhty ... in Karnak, while at the same time, the officials, companions, and chief standard-bearers were tax assessors for the stone statue.[13]

The army was also used as a rudimentary police force, as a means of enforcing laws but also maintaining peace. One letter from a Greek tax collector asks for a couple of soldiers to help with collecting tax from a reluctant villager: 'I have collected the taxes from the residents of the village, all except for Johannes ... he refuses to pay his account ... please send two soldiers to the village where he is being held, because in that way we may be able to get the money that is owed.'[14]

During the internationally peaceful reign of Akhenaten (1350–1334 BCE), there was a heavy military presence at the city of Akhetaten (Amarna) where the soldiers guarded the royal family, protected the boundaries of the town through a series of guard houses, and patrolled paths around the city. It is thought that due to Akhenaten's extreme religious overhaul, he may have felt threatened. As Akhetaten was a brand-new city, none of the soldiers would have originated in the area.

On the lighter side, the military were also drafted in for religious festivals where they would pull the divine barques carrying the gods on procession as well as accompany any river parades. This may have happened as early as the fifth dynasty as there is a block from the temple of Userkaf at Saqqara which shows soldiers accompanying a river procession back from Bastet.[15] The army is also depicted in the Colonnade Hall at Luxor temple, accompanying the sacred barques during the Opet festival from Karnak (FIG 20) to Luxor Temples and back again.

## Officials

The *Report of Wenamun*, from the twentieth dynasty (1185–1070 BCE), crosses over from literary tale to military report, as it seems to record a real event. It is set in the reign of Ramses XI (1098–1070 BCE) who ruled a divided

Egypt with Herihor in the south and Smendes in the north, and records an expedition to acquire Lebanese cedar wood. Wenamun, an Egyptian official started his journey by sailing to Tanis in the Delta before going to the Levant to source wood to repair the boat of Amun which was known as Amun-user-he. He spent nearly a year on the expedition, and relayed encounters with foreign rulers who delayed the mission of Amun for varying reasons, much to Wenamun's annoyance: 'You have let this great god spend these twenty-nine days moored in your harbour. Did you not know that he was here? You are prepared to haggle over the Lebanon with Amun its Lord.'[16]

Although discussing the experience of an expedition outside of Egypt, it must, to a certain extent, be recognisable to the intended audience as being similar to the type of red tape and politics which held up similar expeditions into Egypt. However, the text itself is likely propaganda created to demonstrate that Egypt's reduced global standing was the reason that Wenamun was constantly delayed on what should have been a routine trip.

Other official visits were made by the New Kingdom Viceroys of Kush, who often came from Thebes or Bubastis in the Delta and who were required to work in Aswan as part of their role. They oversaw such projects as canal and irrigation works and accompanied the king on military expeditions. Many of the viceroys left graffiti on cliff walls in the area to show their presence. For example, the viceroy Mermose, from the reign of Amenhotep III (1386–1349 BCE) left graffiti along the Shellal Road in Aswan, which was used as a military route to the south. There was a revolt in the land of Ibhet in the south and an Egyptian military campaign was deployed to quash it. Mermose's presence indicates he may have been part of this campaign.

The viceroy Thure in the reign of Thutmosis I (1524–1518 BCE) wrote on the cliffs of Sehel regarding his visit. 'The third year, first month of Shemu day 22, navigation of His Majesty on this canal in power and strength, upon his return from overthrowing the vile Kush.'[17] The viceroy of Kush also oversaw building works in the region, and opposite a shrine dedicated to Anukis, on the western side of Bibitagong, the viceroy Usersatet left six inscriptions, dating the work to the reign of Amenhotep II (1453–1419 BCE). He added a seventh graffito on an island south of Sahel (Ras Sahel) where he mentions the festival of Anukis which was extended by the king:

> Giving adoration to Amun, doing obeisance to Re-Horakhti by the viceroy, the overseer of the (southern) lands, User(sate)t after coming to see the beauties of Anukis in her beautiful festival of proceeding to Sehel. He made accordingly five canals with workmen scorched in their limbs, this being done anew.[18]

This work would have taken the viceroys away from their home cities and towns and sent them to the remote and unstable desert region of Aswan. Although an important aspect of their role, it shows travelling and living away from home were expected, possible and necessary, otherwise such duties would likely have been filled by local inhabitants.

## Travelling to the Pyramids

Throughout the history of travel in ancient Egypt the pyramids of Giza have been prominent, for their size and for their enduring timelessness. It is very hard to truly comprehend the age of the pyramids, but during the reign of Ramses II the pyramids at Saqqara and Giza were already 1,300 years old and therefore worthy of touristic visits. During this reign, 'a steady stream of day-trippers' visited the pyramids,[19] and left behind graffiti recording their visits:

> Year 47, second month of winter, day 25, the Treasury-scribe Hednakht, son of Tjenro and Tewosret, came to take a stroll and enjoy himself in the west of Memphis, along with his brother Panakht ... He said: O all you gods of the west of Memphis ... and glorified dead ... grant a full lifetime in serving your good pleasure, a goodly burial after a happy old age, like yourself.

However, 1,300 years before this record was left, the builders who were responsible for these structures were also travellers. Excavations and research at the site of Giza have shown many of the workers were corvée labourers.

This all goes back to the annual inundation of the Nile, which meant between June and October, the farmland along the edge of the river was under water. While vital for the fertility of the land, it did mean that farmers were unable to carry out any agricultural work and were unable to earn a living. The king therefore conscripted them to work on monument building for which they were well paid and well fed. Conscription happened locally, with resident officials being told how many people were needed. A twelfth dynasty letter (1991–1782 BCE) from the estate of Senebni explains some of the process:

> the mayor sent me ... to muster the labour force, having charged me ... saying 'as for any persons who you may find missing among them, you are to write to the steward Horemsaf about them' I ... have sent a list of missing persons in writing to the pyramid town.[20]

In the Old Kingdom there was a core group of skilled workmen who laboured on the pyramid all year. It is suggested that peasant farmers worked on a rota system throughout the year as a form of tax to the king to contribute to construction.[21] Those who could not afford a financial donation offered their time and labour instead. They were not treated as slaves. On the contrary, it was an obligation to the state which they carried out when convenient – normally, as stated, through the period of the flood season or through the sowing season when fewer farm hands were needed.[22] However, if they deserted their posts, their families were imprisoned until they returned.

The Great Pyramid is estimated to have taken twenty years to build with a combined workforce of 20,000 people. At any one time there were two gangs of 1,000 workmen, plus their supervisors. These large teams were further divided into five smaller groups of 250 men each, and then manageable teams of about fifty.

The corvée labourers slept in long galleries with room for about fifty, with artisans and skilled workmen living in a town on the eastern edge of the plateau. They ate in a large dining hall, and the bones embedded in the floor shows they had a diet of fish and, more surprisingly, meat – which most of them were unlikely to be able to afford normally.

The material used to build the pyramid also contributed to river miles being put in by workers in the Old Kingdom. For example, granite that was used as casing stone travelled along the Nile from Aswan, whereas white limestone came from the more local Tura quarry between Cairo and Helwan.

## Pilgrimages

Not everyone travelled through Egypt for work and evidence suggests much early travelling had a religious element. There was an itinerary of sites and temples to visit as a means of demonstrating devotion. As one scholar stated, tourism was 'the acquisition of a series of encounters with holy places'.[23] However, due to the evidence being in the form of graffiti or inscriptions on votive offerings, we are presented with a skewed image of the literate element of society. As literacy levels were low in pharaonic Egypt this represents less than five per cent of the population. This has led to the view that prior to the Graeco-Roman period, in regard to travel, 'none took place',[24] which is unlikely to be the case. At the very least, those working in temples would have travelled with festivals and to other temples, and there were always officials travelling the country on diplomatic business.

The fact that those who could leave a note of their presence did so because of the memorable occurrence, does not mean those who were illiterate did not hold the same value to their trips, or not make trips at all. A pilgrimage is any form of spiritual journey, which leads to spiritual growth or understanding in an unknown place, so a trip to a new temple to worship a deity, or a trip on a festival day to tend to the grave of an ancestor are also considered pilgrimages.

Generally, people visited a local temple or shrine dedicated to the god best suited for their needs, but there were national pilgrimage sites which people also tried to visit. A stela in Buto dedicated to the deceased Gmenafharbok describes some of the types of pilgrims to the site. It is dated to the first century BCE and refers to:

> Nobles and servants, embalmers, chiefs of mysteries, people who dispose of offerings in performing the function of priests. O all men who pass on roads, whether they come to the necropolis or pass near this staircase, or go to Imet (Buto) during the day of the full moon to make their oath with the Golden One (Hathor), or to see Wadjet, whose face becomes joyful giving children to pilgrims, or who come from the country of the god, at the period when the plants are green, to worship during the festival of Horus, ... Citizens of Imet and servants, who go and come to kiss the dust in the temple of Silence which has become an antechamber of the tomb ... at the time of famine when the pilgrims descend toward Min and go in procession to him who presides over the country.[25]

The stela mentions three specific pilgrim types: those who come at the time of full moon to address fertility issues, those who come during the festival of Horus and those who appeal to Min during the time of famine. Although this text is particularly late regarding the pharaonic period this idea of visitors to the necropolis is reflected in earlier practices. A common Middle Kingdom funerary monument addressed those who passed by the tomb. For example, the funerary stela of Sehetepibre, found at Abydos states:

> O beloved of the king, O beloved of his city god, priests of Osiris, foremost of those in the west, in Abydos, priesthood of the temple of this god, priests of the king of Upper and Lower Egypt Nimaatre ... and the king of Upper and Lower Egypt Khakaure, justified, their priesthood, those in your city and all the people of Abydos who will pass by this monument going north of south, as your king loves you, as your city gods praise you, as your children remain in your places, as you love life and ignore death, you will say 'a thousand of bread, beer, oxen, fowl, alabaster...'.[26]

These stelae indicate there was a regular presence of people passing by, whether to attend to graves of their own loved ones, to worship the gods of the temple or priests who worked at the site. By appealing to passers-by the stela owner hoped to receive the benefits of a number of voice offerings.

## Aybdos

Abydos was a particularly important site for Egyptians and during the pharaonic period many tomb wall paintings describe and depict a very specific funerary pilgrimage to and from the area. In a funerary context, boats are shown transporting bodies of the deceased to Abydos, the burial place of Osiris. They first appear in the Middle Kingdom (2040–1782 BCE) tombs at Beni Hasan of Amenemhet and Khnumhotep, but then become prominent in Thebes throughout the New Kingdom (1570–1070 BCE). The text accompanying the images outlines the main activities they expected on pilgrimage, including kissing the ground before Osiris, ferrying the god across the river and taking part in the festivals of Osiris.

This ritualistic trip was a means of closely associating the deceased with Osiris and his journey into the afterlife, therefore ensuring their own journeys would also be smooth. There is some debate as to whether this was only a hypothetical journey purely for the realms of funerary art or if, in fact, in life people made the journey to Abydos if they could. It is thought the practice may have originated in the Old Kingdom (2686–2181) with the kings at Buto in Lower Egypt who were taken on funerary journeys to Sais, Busiris, Behbeit-el-Haggar and Mendes.[27] By the time this practice reached the ordinary people it had been condensed to a single trip to Abydos.

Although more prominent in the Middle Kingdom, this funerary journey remained an important feature of funerary art until the Roman period.[28] The Roman mummy mask of Hierax, son of Sarapion from Middle Egypt for example, states: 'May you sail downstream to Busiris, and may you sail upstream to the nome of Abydos, when its resident [Osiris] celebrates the festival of Sokar.'[29]

Although such a ceremonial journey may not have been possible for everyone, there were certainly some people who made the pilgrimage in order to take part in the festival of Osiris, which included a river journey where Osiris was carried from Abydos to Poker (Umm el-Qa'ab), a distance of 1.7 kilometres, and back again. There are records which show active participation with Sehetepibra recorded as taking part in the 'mystery of the Lord of Abydos' and Moses who records carrying Osiris' sacred barque.[30]

Abydos had been an important religious site from the early dynastic period, with the oldest surviving monument dating to the second dynasty (2890–2686 BCE) and built by King Khasekhemwy, but it is likely to have been an important site prior to this. The northern cemetery at Umm el-Qa'ab was situated in front of the mouth of the desert canyon which may therefore have been viewed as the entrance to the west (even though it faces south). This idea is reflected in the Pyramid Texts (PT610:1716–17 PT437: 788–99): 'Betake yourself to the waterway, fare upstream to the Thinite nome, travel about Abydos in this spirit-form of yours which the gods commanded to belong to you. May a stairway to the Netherworld be set up for you to the place where Orion is.'[31] The royal cemetery at Abydos, was on a terrace 10 metres high, to show royal might and superiority.[32]

In general there were two types of pilgrims to Abydos; those who came to participate in the annual festivals of Osiris and leave a commemorative stela, and those who came to Abydos simply to erect a memorial stela.[33]

There was also the unique aspect of worship at Abydos in the Middle Kingdom which was to dedicate a chapel (m'h'.t) or cenotaph, with no associated burial. For many of the owners of these chapels, they are known to have burials elsewhere in Egypt, and the chapel housed a stela dedicated to their deceased selves (even if they were living when it was erected). These chapels varied in size from several metres to just an arch covering the stela, depending on the wealth of the dedicator. The stelae were very similar in nature and content. 'I made this tomb at the terrace of the august god in the sacred land of the western horizon on the ground that gives offerings, the arrival of the place of every god, so that I might be in the following of Wepwawet … going to the terrace of the necropolis following the god in his stride.'[34]

However, regardless of size or how elaborate or simple they were the benefits were the same. These fake burials meant the dedicator received all the benefits of being 'buried' near Osiris. Dedicators also benefited from all the rituals carried out for him, and a common wish was that 'I may inhale incense', which was considered to be the fragrance of the god.[35] The Middle Kingdom votive zone was adjacent to the enclosure which housed the temple of Osiris, just north of Umm el-Qa'ab and was associated closely with birth and rebirth. The chapels all tended to face east where Osiris was reborn.[36] This placement enabled the people named on the stelae to participate and witness all the processions, festivals and rituals to Osiris and therefore to gain benefit from them.[37] 'I have come to this tomb at the terrace of the august god … I shall see Wepwawet and all his festivals at his footsteps.'[38]

More importantly for our study of tourism, m'h'.tw or cenotaphs were placed as official reminders of the pilgrim's visit to the site. They were generally built of mudbrick, and the walls were plastered and whitewashed and may have appeared like 'an army of tents from afar'.[39] None of them were associated with an actual burial and were essentially dummy burials and no funerary items have been found on the site from the Middle Kingdom.[40] They all contained at least one niche, within which was placed a commemorative stela, many of which were carved and commissioned in Thebes or Memphis and then shipped to the site, meaning that not everyone with a stela necessarily visited the site.[41] However, some dedicatees refer to themselves as pilgrims, such as on the twelfth dynasty (1991–1782 BCE) stela of Sehetibre and Ankhu, son of Sehetibre and Sat Montu: 'The honoured Ankhu says: ... I made this stela when I had come to worship the monument of Osiris, Lord of Abydos, Lord of the Sacred Land.'[42]

Ankhu had travelled from the Thinite nome, just north of Abydos, to worship at the site and leave his commemorative stela. The importance of having a stela at Abydos for the spiritual wellbeing of an individual cannot be denied and there is also evidence that shows that some were 'pilgrims by proxy' where they commissioned someone else to erect a stela on their behalf without visiting the site. For example, the keeper of Diadem, Nebupusenwosret during the reign of Amenemhat III (1929–1895 BCE) sent his stela to Abydos with a priest who was attending a jubilee festival for the king: 'This stela fared south with the elder lector-priest Ibi, when the temple priesthood had come to see the king at his beautiful feast of eternity.'[43] In this instance, the owner of the stela was not the traveller to the site of Abydos, but the priest Ibi who had travelled there in a work and spiritual capacity.

In the Third Intermediate Period (1069–525 BCE) actual burials took place at Abydos on the Terrace of the Great God and comprised shaft tombs with a chapel and stela over the top. None of the burials interfered with the older monuments or the second and third dynasty burials at the site indicating that they had been demarcated in some way throughout its long history.[44] By the Roman period stelae were erected at the site as well as cenotaphs, but they seem to be for those who had died young, such as the poet Elemon, who was 20 years old when he died, and a 16-year-old Apollo who died in Alexandria.[45]

## Festivals

There were several important festivals held at Abydos, such as the Middle Kingdom Osiris festival which included a re-enactment of the Osiris myth, including his death and resurrection and the struggle Osiris and his wife Isis had in putting their son Horus on the throne. In line with this festival the votive material from pilgrims shows a shift of motivation in the New

Kingdom (1570–1070 BCE) and rather than focusing on the journey into the afterlife, were more concerned with the site of Abydos as the conception place of Osiris's son, Horus.[46] This was focused around the mammisi, or birth house. At Abydos, the earliest example appears to be from the reign of Nectanebo I, although this was replaced in the Roman period under Emperor Augustus. It was dedicated to Hathor and Ihy the son of Hathor and Horus. However, the dwarf god Bes was also honoured at this temple.

In addition to the festivals at Abydos, discussed above and the oracular processions discussed in chapter 7, there were dozens of religious festivals held in every town in Egypt throughout the year. Some attracted only local people and other larger ones were popular with visitors from other parts of Egypt but also from abroad. Herodotus records six major festivals which included that of Bastet-Artemis at Bubastis which attracted 700,000 worshippers held in the month of Pauni (autumn at the time Herodotus was writing).

The sheer number of people at this festival indicates they came from all over the country and many of them arrived by boat and made lewd gestures at locals on route. Men and women attended the festival, but children were banned, and it is thought this could be a reflection on the activities that went on there.

> The Egyptians meet in solemn assembly not once a year only, but on a number of occasions, the most important and best attended being the festival of Artemis at Bubastis … The procedure at Bubastis is this: they come in barges, men and women together, a great number on each boat; on the way, some of the women keep up a continual clatter with castanets and some of the men play flutes, which the rest, both men and women clap their hands. Whenever they pass a town on the riverbank, they bring the barge close in-shore, some of the women continuing to act as I have said, whilst others shout abuse at the women of the place, or start dancing, or stand up and hitch up their skirts. When they reach Bubastis they celebrate the festival with elaborate sacrifices, and more wine is consumed than during all the rest of the year.[47]

There was dancing and music, as well as military displays, the participants of which probably had to travel to the temple from elsewhere in Egypt. In the Roman period the residents of Oxyrhynchus (near al-Bahnasa, 160 kms southwest of Cairo) travelled 100 kilometres to Antinoopolis regularly to take part in the games which were held during festivals.

Although far later than the pharaonic period, a third century CE record shows a group of flute players and musicians were sent from Oxyrhynchus to another middle Egyptian town for a five-day festival. It is likely when they were not working, they were able to enjoy the revelry and such a practice is likely to have been the same in the earlier periods. The festival was as much about the journey to get there, where the pilgrims stopped en route, as about the festivities once they arrived at the temple.[48]

Some festivals were transient and were not based in one place. They were a great spectacle to attract visitors and basically included a river procession of the god in his sacred barque being transported from one temple to another. The journey could take several days, and the banks of the Nile were thronged with worshippers, pilgrims and tourists. There were a number of these throughout the year and included:

- Ptah of Memphis travelling to Hathor of the Southern Sycamore
- Hathor of Esna visiting Horus of Edfu
- Amun of Thebes travelling to visit his consort Mut

The Beautiful Festival of the Valley was a particularly popular festival in Thebes in the New Kingdom (1570–1070 BCE). The statues of the Theban triad Amun, Mut and Khonsu were carried from Luxor to Karnak Temple along the Sphinx Avenue and then taken by boat to the Theban tombs and Deir el Bahri. The overland festival was private as the Sphinx Avenue was closed to the public, but the boat procession was public. This festival is known from the records from the thirteenth dynasty (1782–1650 BCE) onwards. The public joined the procession on the west bank, often with small painted ancestor busts of their own relatives, before going to the family tomb to celebrate with the deceased. These celebrations could involve a family meal in the chapel above the burial chamber with food left for the ka of the deceased to absorb nourishment from in the afterlife.

The Opet was another festival dedicated to Amun, when the triad of deities from Karnak, Amun, Mut and Knonsu travelled to Luxor Temple and back again. This journey was carried out in an elaborate riverine parade accompanied by priests, singers, dancers, the military and the king. The festival celebrated the union between the king and his queen, which was meant to be represented by that between Amun and his consort Mut. It reaffirmed the legitimacy of the king to rule by associating him and his reign with Amun. The festival lasted up to two weeks and not only was a great celebration for the local inhabitants, but visitors from other parts of Egypt.

The festival where Hathor of Dendera visited her son Horus of Edfu was one of the most important and popular festivals during the Graeco-Roman period. The processions made three stops on the journey: Karnak to visit Mut, the Latpolite nome to gather more pilgrims, and Hierakonpolis, before arriving at Edfu. The journey was 160 kilometres, and it was thought that each stop was at least over one night and enabled more pilgrims to join the procession after each stop. These pilgrims were well taken care of by the local authorities.

> The chief of Nekhen is to furnish 500 loaves of different kinds, 100 jars of beer and 30 shoulders of small farm animals, for the people of the villages, so they pass their journey sated, to drink and enjoy the festival before the venerable god, to anoint themselves with perfume, to play the tambourine with a great noise, with the people of the town (of Edfu).[49]

There was a further eleven days of celebration once the procession arrived at Edfu.

The Khoiak festival was one which commemorated the gathering of the pieces of the murdered Osiris' body by Isis, so she was able to conceive Horus before Osiris went to the underworld. It was popular from the New Kingdom until the Christian period. Part of the rituals included sowing and watering seeds to represent the burial and rebirth of Osiris, as well as the ceremonial raising of the djed pillar representing the spine and stability of Osiris. At Karnak rites were carried out between the eighteenth and twenty-sixth day of the month, with the procession taking place on the final day. According to Herodotus, tens of thousands of pilgrims gathered for it, beating their breasts in mourning for Osiris.

The inauguration of a new Apis bull, in Memphis was also a great attraction for pilgrims as it only happened every twenty- to twenty-five years. According to Herodotus when a new Apis was found it was taken to Nilopolis (75 kilometres from Memphis) for forty days and sailed in procession in a gilded chamber on a boat to the temple of Ptah at Memphis. Once at Memphis more processions were held to ensure the maximum number of worshippers could interact with it.

At the other end of the life of the Apis there was a grand funeral which also attracted thousands of visitors. The rituals associated with the funeral lasted seventy days, the same amount of time as was traditional for human burial as this was the length of time the embalming process took. At the end of the mummification process, the embalmed and wrapped bull was taken in an elaborate procession from Memphis to the Saqqara necropolis. There were two other processions before burial: From the temple of Ptah at Memphis to the lake in the south-west and then from the precinct of the Apis to the Serapeum. These provided ample opportunities for the pilgrims to witness the procession before the burial took place. From the New Kingdom to the reign of Amasis (570–526 BCE) the Apis was placed into a wooden coffin, which then replaced with a stone sarcophagus. Funerals as prominent festival days for the general populous were more popular in the Third Intermediate Period (1069–525 BCE) and Late Period (525–332 BCE).[50]

## Hearing Ear Stelae

Not all the visits to the temples were for public displays of worship, or for public questions to the oracle (chapter 7). Sometimes the pilgrims went to speak privately with the deities, and this saw the introduction of the hearing ear stelae. These allowed the pilgrim to literally hold the ear of the god. These stelae were small, some about the size of a mobile phone, and were inscribed with images of ears, any number between one and more than 300. Most of them were without inscription but some simply bore the name 'Ptah, Lord of Maat' or dedicated to Hathor. It is more common to have the name of the deity on the stela than the name of the donor, and inscriptions may have included things like 'I am calling you, Mut, Lady of Heaven, that you may hear my petitions.'[51]

These stelae have been found in local shrines as well as in homes indicating they were part of a localised religion which had been taken to the more public shrines. The stelae may have represented the 'listening god' or could have provided a vehicle for the words of the pilgrim to go straight to the god who would, hopefully, help them with whatever predicament they had. Stone, metal and wooden ears were also left as votive offerings at the temples of Deir el Bahri, Faras, and Serabit el-Khadim either in gratitude for helping or to persuade the god to hear them.

They were common from the Middle Kingdom, but they were continued to be used well into the Roman period including at the temple of Horus and Sobek at Kom Ombo. The ear stelae were placed in parts of the temple which were accessible to the general public, at a time that the temples were generally closed off. So, for example there was a shrine on the north wall of Amun's temple at Karnak. Such shrines were known as 'Shrine of the Hearing Ear' or the 'place where the God hears petitions'.

These stelae may have been commissioned by individuals but also could have been used by other visitors to the temple. Perhaps the chapels were set up by the priests as a means of bringing more pilgrims to the temples.

## Business of Tourism

By the Ptolemaic and Roman periods many of the temples responsible for larger festivals or international cults were able to make a business out of it, and these soon became important for their continued operation. This need to raise revenue would have been just as important in the earlier periods and, therefore, priests would have facilitated pilgrims as much as possible, whether this was providing ear stelae, or ensuring there was someone on site who could direct prayers to the gods.

Many out-of-town visitors stayed overnight for the festivals, staying, if possible, with friends and family, but this was not always the case and the priests saw opportunities. Thus, by the Ptolemaic Period, temples were quick to offer accommodation for those who were visiting for a festival.

Harmais, a farmer in 157 BCE was travelling from the Heracleopolite nome to the Serapeum in Memphis (about 100 kilometres) and was planning to stay at the Anubeion. This was a structure attached to the Serapeum and was dedicated to Anubis and was rented out to tourists and pilgrims for festivals as an inn or hostelry. Their food and drink was included in the fee. Similar inns existed at the Arsenuphis near the temple of Philae, and the Aphrodision in Memphis. It is likely that such inns also existed in most of the main temples of Egypt during the pharaonic period which attracted visitors for major festivals, although perhaps on a more ad hoc basis.

Many priests were rather entrepreneurial, and they made sure that the pilgrims were able to spend at the festivals by making available food and drink as well as souvenirs, offerings and trinkets. Amulets and souvenirs varied from temple to temple, with a Hathor headed sistra, dedicated to the goddess of Hathor at Dendera or Deir el Bahri, to a cat mummy at Bubastis. The earliest individual examples of cat mummies go back to the reign of Amenhotep III (1386–1349 BCE) but the large-scale deposits of animal mummies date to the reign of Nectanebo II in the thirtieth dynasty (360–343 BCE) onwards indicating an increase in demand.

There is much misrepresentation of these animal mummies with people stating that the thousands of mummified cats shows how much the Egyptians loved cats. Not so. Greek sources also claim that it was a capital offence to harm a cat as they were considered so sacred. This is completely at odds with the practice of cat slaughter for mummification. Herodotus records that: 'Cats which have died are delivered to Bubastis, where they are embalmed and buried in sacred receptacles.'[52] This indicates that pilgrims brought animal corpses themselves to offer to deities, whereas the sheer number and manner of the animals' deaths indicates many were raised at sanctuaries and pilgrims simply paid for animals to be dedicated without a thought for whether they died of natural causes – which reconciled the idea of the animal as sacred with their desire to dedicate a mummy.[53]

Animal mummies as dedications were a money spinner, and the cats were bred purely for slaughter to sell to the pilgrims. X-rays of mummies have shown a number of them were young kittens whose necks had been broken which is very different from cats living out their lives in luxury. It is the same with the ibis and the crocodile mummies, although these were closer to fraud as the mummies themselves often comprised a combination of debris rather than a mummified animal. These mummies have been found in their hundreds of thousands, and in the nineteenth century they were so numerous that many were shipped to Europe to be used as fuel.[54]

This fraud was known at the time and a priest of Isis at the ibis necropolis at Saqqara, during the reign of Ptolemy VI Philometor (180–145 BCE), records that it should be, 'one god in one vessel', indicating that each vessel bought by a pilgrim should contain one bird. Excavations of many of the vessels have shown some were empty, and some were a jumble of bones indicating either that demand outstripped supply, or the priests were simply making the most of the resources they had and knew the tourists would not suspect anyway.

Even with only a limited number of tourists at sites in Egypt during the pharaonic period, out of village visitors would also have been welcomed as they increased the sale of food, drink and accommodation. This is the start of the tourist industry. Albeit on a very small scale.

## Epilogue

# The Future of Tourism in Egypt

Egypt has had a long history of tourism, from internal tourism in the pharaonic period to international tourism starting in the Ptolemaic period – a history spanning 5,000 years. However, in that time, although some of the sites have remained the same with the pyramids being popular throughout the eras, the nature of tourism has changed. It has spanned from practical, to religious, to sightseeing, to mass tourism, to check-list tourism. This constant evolution of the sites has ensured that there will always be people who want to visit. For example, the mortuary temple of Amenhotep III (1386–1349 BCE) at Hawara was reinvented as the Labyrinth by Herodotus, the temple of Hatshepsut (1498–1483 BCE) at Deir el Bahri started as a mortuary temple and then became a site of healing first by Imhotep, a deified architect, then Asclepius, and Amenhotep son of Hapu, another architect. This perpetual reinvention keeps the tourists coming. For a couple of centuries now, as visitors are no longer worshipping at the sites, their current incarnations are as historical monuments, and as representations of the histories they have lived. This will, therefore, only interest those with a passion for history, art and culture.

Every tourist has brought something to Egypt, but many have also taken something away which, as the numbers have increased, has become more detrimental to the monuments and culture of Egypt. Since the Arab Spring in 2011 the Egyptian tourism industry has seen a continual decline, and every time it seems to make a come-back, another national or international crisis happens which once more reduces tourist numbers.

There is no doubt that tourism is a powerful economic driver for the Egyptian people, but it is also unstable and can be affected by civil unrest and terrorism, and whilst for the tourist it is merely a change of location for their holiday, for the country it can be catastrophic. Not just for the economy, but also creating a general feeling of instability, meaning the country is less desirable for investment, reducing employment opportunities and creating a generation of nihilistic youth.[1] In 2010, when Egypt had a reached touristic high point, nearly seven per cent of the workforce of Egypt was directly employed in the tourism industry.[2]

Following the Arab Spring when the Egyptian government and President Mubarak were overthrown, tourist arrivals declined by a third compared to 2010. Prior to the 2011 revolution, tourism was at a high of about 15 million visitors a year, which plummeted to just nine million in 2014, and less than nine million in 2015.[3] Between 2010 and 2014 tourism revenues had fallen between 40 per cent and 54.34 per cent.[4] Add this decline to an already weak economy when, on 1 January, 2011, Egypt's internal debt was $164 billion, or 75 per cent of its GDP, and unemployment was as high as 27.4 per cent in 2010, putting Egypt in a weak position.[5]

However, tourists have short memories, and the numbers were increasing in 2012 – although they were definitely fewer than 2010. Just when the industry looked to be improving, President Mohammed Mursi was overthrown in a military coup in 2013, adding to the image of political instability which repelled tourists with many cancelling or postponing their trips. The Minister of Tourism at the time, Hisham Zazou commented '2013 was the worst year on record for Egypt's tourism industry.'[6]

However, the reduction of tourists since the 2011 revolution was still being felt in 2015 and whilst tourism in general increased globally, it had not really picked up in Egypt to match the pre-2011 levels.[7] There was a wider knock-on effect to the economy due to the revolution which has seen the foreign currency reserves at less than half of what they were before January 2011 ($16bn-$17bn) which threatened the country's ability to buy fuel and food. The lowest contingency level is $15bn. This is due to the main source of foreign currency being tourism.

One aspect of the revolution which perhaps people had not considered was the impact on the ancient monuments and, thus, the world's cultural heritage. Shortly after the revolution in 2011, many sites in Egypt were looted, from museums including the Cairo Museum, to archaeological sites. 'There is not a single site in Egypt that has not suffered the attack of the land mafia and looters, from the necropolis of the New Kingdom at Aswan to areas of the Eastern Desert along the Red Sea.'[8]

With little to no security at many sites, especially those out of the way places, many artefacts were lost forever, and many archaeological sites were damaged beyond repair. The Mallawi Museum for example had about 1,089 objects in the collection. Approximately 1,050 were destroyed or looted and lost forever.

One thing that does need to be considered is the sheer size of some of these archaeological areas – for example, the distance between Giza to Dahshur is about 23 kilometres long and about 14 kilometres wide through desert areas. This made it hard for security to cover, especially when there were other things going on, and many looters were armed with automatic weapons from Libya, the type the security services could only dream of.⁹ Add to this, the fact that the Morsi government did not care about the monuments and had 'an ideological tendency to de-legitimise the pre-Islamic past as an essential component of national identity'.¹⁰ In short, there was little effort made to stop the looting.

Many of the items stolen passed through the port of Ain Sokhna south of the Suez on the west bank of the Red Sea, which had no antiquities control until November 2012. The Ministry of Antiquities' Recovery and Repatriation Unit was able to recoup thousands of stolen artefacts in Egypt and the wider world through a website tracking database.¹¹ Most of the looting had slowed down by 2014 as security was once more increased at vulnerable sites. Groups of heritage activists did what they could to protect the monuments as well.

Sarah Parcak and her digital imaging team did a great deal of work in identifying the extent of looting at archaeological sites including at Saqqara, Dashur, Lisht, and el Hibeh. Damage was caused by illegal digging for artefacts that comprised burial shafts, some three to four metres deep with the entrances strewn with bandages, human remains and pottery. There was also large-scale bulldozing of areas to illegally construct cemeteries.¹² As time progressed the number of illegal pits increased exponentially. For example, the Amenemhat III pyramid complex had 50 pits in May 2011, 988 in September 2012, and 1,159 pits by March 2013, as well as an illegal cemetery that was constructed.¹³ These illegal excavations may have revealed more than 1,000 previously unknown (and possibly undisturbed) tombs at the site of Lisht alone, and with no idea of what was stolen, the archaeological and historical significance is lost for good. At the village of Dayr Abu Hinnis, a Christian village with three orthodox and two evangelical churches, a third dynasty (2686–2613 BCE) cemetery was bulldozed to make room for pilgrims heading to the church.¹⁴ Similar land grabbing problems occurred at Amarna, the capital city of Akhenaten, destroying much of the archaeology there.

Sadly, it was not only the ancient monuments which were targeted but also Islamic monuments in medieval Cairo, including pulpits, prayer niches pointing towards Mecca, lanterns, carved ceilings, embroidered fabrics and other portable items. Entire buildings were also sold and destroyed despite laws protecting them from such destruction. With no one to enforce these laws, they were pointless.

Although some of the looting had an organised crime element to it, much was likely to be opportunistic and seen as ways to make quick money. Many operations were run by a local leader, who controlled gangs of children digging for artefacts, and then passed the finds on to low-level antiquities dealers. Parcak believes that creating opportunities for local people or increasing tourism numbers to the area would prevent this type of activity.[15] Monica Hanna adds that many tourist sites do not actively benefit local communities so there is little incentive for the locals to protect monuments.[16]

If this deliberate destruction were not bad enough, neglect has added to the damage with the site of the Ramesside temple of Ra at Heliopolis being flooded with solid waste, resulting in vegetation growing out of the mudbrick walls. The local mafia also tried to hide archaeological features by setting fire to the site more than once.[17]

## The Bounce Back

Whilst it is possible for some countries to bounce back quickly following violent conflicts or terrorist attacks, doing so depends greatly on the stability of the country to start with.[18] Taking into consideration the regular terrorist attacks throughout the 1990s, there has not been the long-term appearance of stability in Egypt that many tourists need to feel safe. As R. Butler and W. Suntikul state: 'It is hardly possible to describe North Africa and the Middle East, in particular countries such as Egypt, Israel, Libya or Tunisia as entirely peaceful regions even if no formal warfare is taking place.'[19] Konstantinos Tomazos takes it a little further and describes Egypt as: 'a zone of moderate conflict and political instability with sporadic violent acts and terrorist attacks throughout the country to be taken into consideration'.[20]

One of the main side effects of the 2011 revolution was a general breakdown in security, with border security between Libya and Egypt and the whole Western Desert deteriorating, meaning they were no longer safe for tourists to visit. Even tourist resorts like Luxor which, prior to 2011, was swarming with the white-clad tourism police, security was suddenly conspicuous in their absence. It has even been said that:

> Political and social turmoil brought about by Arab Spring, created fertile conditions for these organisations [Muslim Brotherhood, Islamic State of Iraq and Syria (ISIS)] to rise to prominence leading to dramatic transformations in the security environment of the country and region now.[21]

Following the Arab Spring in 2011, the Muslim Brotherhood leader Mohammed Morsi became president until there was a military coup d'état which led to the rise of another military leader, Abdel Fattah al-Sisi in 2014. The Muslim Brotherhood was then declared a terrorist organisation, and

Morsi was sentenced to death. There seemed to be one politically volatile event after another between 2011 and 2014.

But with every rise in tourism numbers following one of these political setbacks, the industry was once more halted by another set of incidents. There were continued terrorist threats when on 10 June 2015 ISIS Sinai Province staged a failed suicide bomb attack on Karnak Temple, and the Italian Consulate in Cairo was bombed on 11 July 2015.[22] In the same year, a Croatian engineer was kidnapped just outside Cairo and beheaded and in early 2016, Italian student Giulio Regeni was kidnapped, tortured and killed, apparently by Egyptian security services.[23] It did not help that official Egyptian propaganda contradicted facts concerning the events, damaging any trust the world had in their claims of safety.

A plane crash in Sinai in 2015 also raised additional concerns about safety which was followed by an Egyptian military airstrike on a group of Mexican tourists who were mistaken for insurgents in the Whahat area of the Western Desert. Twelve people were killed.[24] The authorities claimed the tourists were in a restricted area, but a local Egyptian claimed this was not the case and that they even had a police escort, which is likely considering the danger of entering the desert areas.

Therefore, unsurprisingly, in 2016 tourist figures were down by 40 per cent on 2015. In 2016 figures dropped again by 60 per cent in one quarter following the Russian Metrojet crash from Sharm el Sheikh which killed 224 passengers, and which was claimed to be an ISIS attack. Flights from Russia were grounded by the impact, having a knock-on effect as other countries advised against travelling to Egypt, including those from the Middle East, with tourist numbers declining by a further 28.6 per cent.[25] This terrorist attack decimated the industry in the Red Sea, with tourist occupancy rates falling by approximately 90 per cent in Sharm el Sheikh and 80 per cent in Hurghada. Many hotels closed which resulted in numerous job losses. Sisi pledged that he would deal with the terrorist threats from ISIS Sinai Province and bring much needed stability to Egypt.

The lack of visitors cannot be blamed entirely on the fear of the individuals themselves but also the tourist industry infrastructure. When governments issue warnings about travel to a particular country, insurance premiums increase dramatically. Some do not even cover travel to Egypt and this can often be the deciding factor for tourists who just want a relaxing holiday.

From 2016 onwards, there was a massive PR campaign to show that Egyptian airports were safe, and that tourists would have a wonderful time in the land of the pharaohs. This included getting celebrities to endorse the country, including Morgan Freeman on the #thisisEgypt.com website. The hashtag itself was hijacked on social media following the kidnapping and murder of Giulio Regeni to highlight Egypt's human rights issues. But the Egyptian authorities kept at it, including the Central Bank of Egypt (CBE) deferring any debts owed by the tourism sector for three years to try to kick-start the economy.[26] People were optimistic about the economy and believed 'Egypt now has the chance to re-emerge as a major economic power in both the region and the world.'[27]

However, this all fell almost into insignificance when the Covid pandemic hit in 2020 and airports the world over were closed, flights grounded, and entire countries placed under lockdown conditions. Even local tourism was not possible which decimated the Egyptian economy further. However, as the world opens, it is safe to say that the tourist industry in Egypt will bounce back, regardless of what the future throws at it, because of two main factors. One, Egypt owns one third of all ancient monuments in the world and two, the Ministry of Tourism has the ability to compartmentalise tourist 'zones'. This compartmentalisation means areas like Cairo and Luxor are tourist bubbles, but travelling to Middle Egypt, for example, or Abu Simbel, by road requires going with an armed convoy to give the impression of safety. Therefore an element of safety can always be presented by keeping tourists in these highly controlled areas.

As Jalāl al-Dīn al-Suyūṭī (1145–1505 CE) summed up: 'The Pyramids situated in the provinces of Egypt are the kind of buildings which do not perish, while time (itself) perishes, and while the guideposts of time disappear, the fame of the pyramids do not.'[28]

The pyramids have stood for 5,000 years and are likely to stand for another 5,000 years.

Although as we have seen attitudes towards, and interest in seeing them change over the millennia, they have never waned. No doubt interest will continue to transform over the next five millennia, but that flame of inspiration will never be extinguished.

# Notes

**Introduction**
1. https://dictionary.cambridge.org/dictionary/english/tourist (accessed 16 February 2023).
2. Foertmeyer 1989:1.
3. Beness & Hillard 2003:204.
4. Mandler 1999:126.
5. Francaviglia 2011:70.
6. Mandler 1999:129.
7. Mairs & Muratov 2015:12.
8. Mairs & Muratov 2015:12.
9. Gange 2015:83.
10. Thompson 2016:229.
11. Thompson 2016:262.
12. Navrátilová 2009:20.
13. Mairs & Muratov 2015:15.
14. Buzard 1993:4.
15. Usick 2002:16.
16. Partridge 1996:80.
17. Partridge 1996:98.
18. Partridge 1996:104.
19. Partridge 1996:15.
20. Partridge 1996:14.
21. Partridge 1996:23.
22. Bagnall & Cribiore 2006:371.
23. Ibrahim & Ibrahim 2011:50.

**Chapter 1**
1. Slyomovics 1989:144.
2. Ibrahim & Ibrahim 2011:50.
3. Richter & Steiner 2008:939.
4. Richter & Steiner 2008:952.
5. Ibrahim & Ibrahim 2011:52.
6. https://countryeconomy.com/demography/population/egypt?year=1990 (accessed 26 October 2022).
7. El-Kholei & Abu-Zekry 1998:42.
8. Ibrahim & Ibrahim 2011:55.
9. Gray 1998:96.
10. Zallio 2010:1.
11. Gray 1998:109.
12. Mitchell 1995:9.
13. Mitchell 1995:9.
14. Wynne-Hughes 2012:625.
15. Zahran 2000:185.
16. Abu-Zekhry & El Kholei 1998:1.
17. Morcos et al 2000:36.
18. El Aref 2006.

19. El Aref 2006.
20. Morcos et el 2003:Follow-up activities.
21. Zahran 2000:188.
22. El Aref 2006.
23. El Aref 2006.
24. El Aref 2006.
25. Morcos et el 2003: Concluding remarks.
26. Abu-Zekry & El-Kholei 1998:4.
27. Gray 1998:91.
28. Gray 1998:94.
29. Richter & Steiner 2008:949, and Gray 1998:94.
30. Kuppinger 2005:353.
31. Al-Ahram weekly, 21–26 May 1998 quoted in Kuppinger 2005:354.
32. http://guardians.net/hawass/conservation2.htm (accessed 16 February 2022).
33. Kuppinger 2005:349.
34. Wynne-Hughes 2012:629.
35. Mitchell 1995:9.
36. Mitchell 1995:9.
37. Abu-Zekhry & El-Kholai 1998:5.
38. El-Bastawissi 2000:164.
39. Kuppinger 2005:353.
40. https://guardians.net/hawass/conservation.htm (accessed 9 February 2023).
41. Wynne-Hughes 2012:636.
42. www.factumfoundation.org/pag/1015/scanning-seti-the-regeneration-of-a-pharaonic-tomb
43. https://www.newyorker.com/magazine/2016/11/28/the-factory-of-fakes (accessed 14 August 2022).
44. https://dailynewsegypt.com/2008/01/01/el-gourna-villagers-face-uncertain-future-as-evacuation-continues/ (accessed 8 February 2023).
45. Wynne-Hughes 2012:628.
46. https://www.thestar.com/news/2007/04/30/pyramids_of_giza_in_peril.html (accessed 12 August 2022).
47. https://investigations.peta.org/egypt-working-animals/ (accessed 12 August 2022).
48. Abu-Zekhry & El-Kholei 1998:4.
49. Abu-Zekry T. & El-Kholei A.1998:6.
50. https://egyptindependent.com/egypt-implements-its-first-ever-accessible-pathway-for-the-disabled-in-karnak-temple (accessed 13 August 2022).
51. Kuppinger 2005:362.
52. Kuppinger 2005:362.
53. Kuppinger 2005:363.
54. Kuppinger 2005:367.
55. Humphreys 2021:161.
56. Richter & Steiner 2008:946–8.
57. Richter & Steiner 2008:948.
58. Kuppinger 2005:354.
59. http://www.yasminaofcairo.com/dancersb.htm (accessed 26 August 2022).
60. https://journeythroughegypt.com/about-us/ (accessed 26 August 2022).
61. Abdulrahman 2021.
62. El Gawhary 1995:27.
63. Abdulrahman 2021.
64. https://www.globotreks.com/destinations/egypt/garbage-city-cairo/ (accessed 10 October 2022).
65. https://www.viator.com/tours/Cairo/Cave-Church-Garbage-City-and-The-City-of-the-Dead-In-Cairo/d782-72617P24 (accessed 10 October 2022).
66. https://www.manchestereveningnews.co.uk/news/uk-news/he-wants-me-not-house-17633573 (accessed 20 November 2022).

Notes 187

67. Mitchell 1995:9.
68. https://www.manchestereveningnews.co.uk/news/uk-news/he-wants-me-not-house-17633573 (accessed 20 November 2022).
69. https://egyptianstreets.com/2019/09/03/he-used-me-as-a-bank-stories-of-foreign-women-married-to-egyptian-men/ (accessed 20 November 2022).
70. Personal correspondence with Luxor brides.
71. https://www.manchestereveningnews.co.uk/news/uk-news/iris-82-toyboy-husband-mohamed-23666243 (accessed 20 November 2022).
72. https://5pillarsuk.com/2014/01/30/more-egyptian-men-marry-older-foreign-women/ (accessed 20 November 2022).
73. https://5pillarsuk.com/2014/01/30/more-egyptian-men-marry-older-foreign-women/ (accessed 20 November 2022).
74. Quoted in Behbehanian 2000:33.
75. Behbehanian 2000:34.
76. Behbehanian 2000:33.
77. Behbehanian 2000:34.
78. El-Gawhary 1995:26.
79. El Gawhary 1995:27.
80. El Gawhary 1995:27.

**Chapter 2**

1. Humphreys 2021:25.
2. https://www.business-live.co.uk/retail-consumer/thomas-cook-returns-online-travel-18941413 (accessed 21 November 2022).
3. Edwards 1878: 18.
4. Tyldesley 2005:103.
5. Published by Thomas Cook.
6. Budge 1906:22.
7. Edwards 1878:136.
8. Flaubert 1979:160 & 54.
9. Budge 1906:32.
10. American journalist William Cowper Prime, 1855–1856, in Humphreys 2021:36.
11. 1906:31–2 & 24–5.
12. Budge 1906:27.
13. Hankey 2007:76.
14. Booth 2018:226.
15. Hamed 2013:435.
16. Jones 2003:258.
17. Humphreys 2011:101–2.
18. Jackson & Stamp 2002:173–4.
19. Humphreys 2011:105.
20. Humphreys 2011:105.
21. Humphreys 2011:197.
22. Kelly 1902:69.
23. https://www.peta.org.uk/blog/animal-rides-bangiza-pyramids-egypt/ (accessed 16 February 2023).
24. Kelly 1902:76–7.
25. Budge 1906:13.
26. Humphrey 2021:39.
27. Edwards1878:5.
28. Humphreys 2021:36.
29. Humphreys 2021:33.
30. Brendon 1991:121.
31. British artist William Henry Bartlett in Humphreys 2021:34.
32. Hunter 2004:29.
33. Hunter 2004:32.

34. Humphreys 2021:9.
35. Budge 1906:2.
36. Budge 1906:1.
37. Humphreys 2021:17.
38. Brendon 1991:125.
39. Humphreys 2021:12.
40. Manley 1991:164.
41. Budge 1906:20.
42. Manley 1991:14.
43. Brendon 1991:124.
44. M.L.M. Carey in Humphreys 2021:43.
45. Brendon 1991:126.
46. Hamilton 2005:181.
47. Brendon 1991:200.
48. Manley 1991:171.
49. Keck 2010:299.
50. Captain John Ardagh in Brendon 1991:135–6.
51. Hunter 2004:46.
52. Boorstin 1961.
53. Keck 2010:295.
54. Romer 1981:125.
55. Keck 2010:299.
56. Hunter 2004:42.
57. Brendon 1991: 126–7.
58. Keck 2010:295.
59. Hunter 2004:41.
60. Hamilton 2005:179.
61. Hamilton 2005:180.
62. Humphreys 2011:76.
63. Hamilton 2005:169.
64. Hopwood 1987.
65. Edwin de Leon in Humphreys 2011:77.
66. Humphreys 2011:76.
67. Humphreys 2011:75.
68. Manley 1991:75.
69. Edwards 1978:17.
70. Humphreys 2011:94.
71. Life Magazine 11 February 1952 in Humphreys 2011:98.
72. Kuppinger 2005:358.
73. Humphreys 2011:107.
74. Kuppinger 2005:358.
75. Humphreys 2011:108.
76. Egyptian Gazette, 15 January 1902 quoted in Kuppinger 2005:358.
77. Humphreys 2011:109.
78. Kuppinger 2005:366.
79. Kuppinger 2005:359.
80. https://www.marriott.co.uk/hotels/travel/caimn-marriott-mena-house-cairo/ (accessed 23 November 2022).
81. Hunter 2004:48.
82. Hunter 2004:49.
83. Pierre Loti, 1910 in Manley 1991:170 & Humphreys 2011:182.
84. Hunter 2004:32.
85. Humphreys 2011:182.
86. Guglielmi 2001:365.
87. In 1860 Rev. Henry Moule introduced the dry earth closet, which comprised dry earth and ash being placed in a bucket after it had been used.

88. Berkefeld filters were introduced in 1891, where water was passed through earth.
89. Manley 1991:208.
90. Humphreys 2011:199.
91. Humphreys 2011:203.
92. Budge 1906:42.
93. Hunter 2004:43.
94. Budge 1906:36.
95. Budge 1906:43.
96. Keck 2010:296.
97. Brendon 1991:136.
98. Carnarvon 2009:8.
99. Brendon 1991:133.
100. Budge 1906:22.
101. Romer 1982:127.
102. Humphreys 2021:19.
103. Keck 2010:298.
104. Douglas Sladen 1856–1947 in Humphreys 2011:88–9.
105. 1902:102.
106. Chapter 1.
107. Booth 2018: 186.

**Chapter 3**
1. Anderson & Fawsy 1987:7.
2. Neret, 2002:12.
3. Tyldesley 2005:45.
4. Bednarski 2005:9, Neret 2002:12 and Tyldesley 2005:47.
5. Anderson & Fawsy 1987:7.
6. Bednarski 2005:13.
7. Russell 2001.
8. Vivant Denon 1799 in Manley 1991:167.
9. Manley 1991:184.
10. Tyldesley 2005:49.
11. This won't be discussed in this book as it has been extensively covered elsewhere, and for the most part took place in England and in France so not directly related to tourism at this point.
12. Pico della Mirandola 1463–94 quoted in Thompson 2015:60.
13. Romer 1981:124.
14. Morecroft 2018:50.
15. Tyldesley 2005:41.
16. Tyldesley 2005:45.
17. Thompson 2015:65.
18. Thompson 2015:66.
19. Thompson 2015:68.
20. Quoted in Tyldesley 2005:73.
21. Tyldesley 2005:92.
22. Macquitty 1965:125.
23. Tyldesley 2005:83.
24. Tyldesley 2005:98.
25. Quoted in Tyldesley 2005:77.
26. The numbering of the tombs relates to the order they were discovered in and was introduced by John Gardiner Wilkinson between 1821 and 1833.
27. Quoted in Tyldesley 2005:93.
28. Tyldesley 2005:103.
29. Reeves & Wilkinson 1996:60.
30. Thompson 2015:67.
31. Thompson 2015:66–7.

32. Baines & Malek 1984:24–5.
33. Reeves & Wilkinson 1996:52.
34. Quoted in Romer 1981:32.
35. Romer 1981:33.
36. Bruce 1790.
37. Tyldesley 2005:44.
38. Romer 1981:34.
39. Morecroft 2018:46.
40. Manley 1991:102.
41. James Bruce 1768 in Manley 1991:123–4.
42. Manley 1991:168.
43. Morecroft 2018:43.
44. These are now in Turin – ASUT 1763 a & b.
45. Manley 1991:80.
46. Francaviglia 2011:69.
47. Manley 1991:51.

**Chapter 4**
1. Curran 2003:102.
2. Tyldesley 2005:40.
3. Timbie 1998:423 and Grossman 1998:289.
4. Frankfurter 1998:38.
5. Haycock 2003:134.
6. Dannenfeldt 1959:10.
7. Dannenfeldt 1959:12.
8. Burnett 2003:66.
9. Bagnall & Rathbone 2004:108.
10. Timbie 1998:423.
11. Meinardus 2002:84.
12. Frankfurter 1998:18.
13. Meinardus 2002:75.
14. Meinardus 2002:91.
15. Davis 1998:326.
16. Frankfurter 1998:17.
17. Behlmer 1998:355.
18. Quoted in Behlmer 1998:367.
19. Behlmer 1998:350.
20. Behlmer 1998:345.
21. Behlmer 1998:362.
22. Timbie 1998.
23. Frankfurter 2006:435.
24. Life of John the Little, chapter 75, in Mikhail & Vivian 1997:48.
25. Life of John the Little, chapter 75, in Mikhail & Vivian 1997:48.
26. Life of John the Little, chapter 75, in Mikhail & Vivian 1997:49–50.
27. Frankfurter 2006:445.
28. Meinardus 2002:68.
29. Behlmer 1998:361.
30. Gabra 1993:123–24.
31. Montserrat 1998:269.
32. Montserrat 1998:271.
33. Montserrat 1998:275.
34. Montserrat 1998:273.
35. Meinardus 2002:40.
36. Montserrat 1998:267.
37. Bagnall & Rathbone 2004:119.
38. Frankfurter 1998:34.

39. Montserrat 1998:272.
40. Meinardus 2002:69.
41. Grossman 1998:287.
42. Davis 1998:314.
43. Grossman 1998:290.
44. David 1998:306.
45. Davis 1998:315.
46. Bagnall & Rathbone 2004:110.
47. Frankfurter 1998:3.
48. Dannenfeldt 1959:14.
49. Thompson 2015:54.
50. Broadhurst 1952:50.
51. Burnett 2003:68.

**Chapter 5**
1. Tyldesley 2005:35.
2. Thompson 2015:44.
3. Quoted in El Daly, 2003a:12.
4. Burnett 2003:76.
5. Burnett 2003:78–9.
6. El Daly 2005:49.
7. Nemoy 1939:22.
8. El Daly 2005:32.
9. El Daly 2005:35.
10. El Daly 2005:33.
11. El Daly 2005:39.
12. Tyldesley 2005:37.
13. Al Maghrabi in El Daly 2005:36.
14. El Daly 2005:44.
15. El Daly 2005:38.
16. El Daly 2003:44.
17. El Daly 2005:37.
18. Rodenbeck 1988:40.
19. Paraphrased by Howard Vyse in Thompson 2015:50.
20. El Daly 2005:37.
21. El Daly 2005:31.
22. El Daly 2005:35.
23. El Daly 2005:34.
24. Thompson 2015:50.
25. Broadhurst 1952:44.
26. El Daly 2005:38.
27. El Daly 2005:41–2.
28. El Daly 2005:54.
29. El Daly 2005:55.
30. Thompson 2015:70.
31. Manley 1991:48–9.
32. Broadhurst 1952:33.
33. El Daly 2005:5.
34. Manley 1991:48–9.
35. Broadhurst 1952:36.
36. Broadhurst 1952:37.
37. Manley 1991:66.
38. El Daly 2003:45.
39. El Daly 2005:50.
40. Broadhurst 1952:54.

41. Quoted in Thompson 2015:45–6.
42. El Daly 2005:52.
43. Nemoy L. 1939:21.
44. Thompson 2015:48.
45. Quoted in Thompson 2015:49.
46. Quoted in Thompson 2015:47.
47. El Daly 2005:49.
48. Thompson 2015:58.
49. Burnett 2003:68.
50. Quoted in Thompson 2015:70.
51. Thompson 2015:47.
52. Thompson 2015:48.
53. Nemoy 1939:28 and El Daly 2003a:54.
54. El Daly 2003a:54.
55. Broadhurst 1952:46.
56. Thompson 2015:47.
57. Dominican friar Felix Fabri in 1483 from Curran 2003:103.
58. El Daly 2003a:55.
59. El Daly 2003a:55.
60. El Daly 2003a:57.
61. Broadhurst 1952:57.
62. Al-Maqrizi in El Daly 2003:159.
63. El Daly 2003a:65.
64. Thompson, 2015:48.
65. El Daly 2005:40.
66. El Daly 2003a:58.
67. Quoted in Thompson 2015:43.
68. El Daly 2005:42.

**Chapter 6**
1. Foertmeyer 1989:168.
2. Momigliano 1966:129.
3. Quoted in Tyldesley 2005:29.
4. Herodotus 1972:131.
5. Manley 1991:167.
6. Plutarch 1973: 281–2.
7. Morcos 2000:40.
8. Lloyd 2000:395.
9. Foertmeyer 1989:281 & n.5.
10. Foertmeyer 1989:281 & n.5.
11. Foertmeyer 1989:283.
12. Grafton Milne 1916:78.
13. Bagnall & Rathbone 2004:48.
14. Foertmeyer 1989:264.
15. Foertmeyer 1989:160.
16. Grafton Milne 1916:77.
17. Bagnall & Cribiore 2006:373.
18. Bagnall & Cribiore 2006:366.
19. Bagnall & Cribiore 2006:365.
20. Bagnall & Cribiore 2006:367.
21. Foertmeyer 1989:68.
22. Foertmeyer 1989:77.
23. Foertmeyer 1989:78.
24. Foertmeyer 1989:209.
25. Foertmeyer 1989:209.

26. Frankfurter 1998a:219.
27. Strabo 17.1.27.
28. Tyldesley 2009:99.
29. Grafton Milne 1916:78.
30. Grafton Milne 1916:77.
31. Foertmeyer 1989:117.
32. Foertmeyer 1989:169.
33. Foertmeyer 1989:11.
34. Foertmeyer 1989:12.
35. Foertmeyer 1989:27.
36. Foertmeyer 1989:259.
37. Chauveau 2002:28.
38. Chauveau 2002:33.
39. Quoted in Tyldesley 2009:100.
40. Tyldesley 2009:99–100.
41. Bagnall & Rathbone 2004:49.
42. Dio 51.16.5.
43. Tyldesley 2009:203.
44. Suetonius, The Twelve Caesars 18:1.
45. Tyldesley 2009:73–4.
46. Strabo XVII, 8.
47. Maehler 2004:5.
48. John of Nikiu quoted in McKenzie 2004:109.
49. Tyldesley 2009:88.
50. Barnes 2004:71.
51. Brazil 2000:52.
52. Jensen 1990:81.
53. Barnes 2004:74.
54. Brazil 2000:49.
55. Vallance 2000:95.
56. Grafton Milne 1916:77.
57. Manley 1991:79.
58. Herodotus 1972:179.
59. Foertmeyer 1989:16.
60. Foertmeyer 1989:17.
61. Foertmeyer 1989:18.
62. Foertmeyer 1989:108.
63. Rutherford 2003:181.
64. Rutherford 2003:179.
65. Rutherford 2003:188.
66. Rutherford 2003:182.
67. Rutherford 2003:177.
68. Rutherford 2003:179.
69. Foertmeyer 1989:260.
70. Grafton Milne 1916:80.
71. Foertmeyer 1989:24.
72. Foertmeyer 1989:75.
73. Tyldesley 2005:31.
74. Romer 1981:29.
75. Foertmeyer 1989:30.
76. Foertmeyer 1989:70.
77. Foertmeyer 1989:29.
78. Foertmeyer 1989:82.
79. Grafton Milne 1916:80.
80. Foertmeyer 1989:68.

81. Romer 1981:31.
82. Romer 1981:30.
83. Tyldesley 2005:34.
84. Romer 1981:31.
85. Foertmeyer 1989:70.
86. Foertmeyer 1989:35.
87. Beness & Hillard 2003:204–5.
88. Rutherford 1998:234.
89. Rutherford 1998:243.
90. Rutherford 1998:247.
91. Rutherford 1998:255.
92. Foertmeyer 1989:37.
93. Foertmeyer 1989:73.
94. Foertmeyer 1989:36.
95. Foertmeyer 1989:38.

**Chapter 7**
1. McDowell 1990:113.
2. McDowell 1999:109.
3. Ripat 2006:307.
4. A papyrus in the Brooklyn Museum depicts the shrine of Amun-Re being carried towards Pemou and the priests. Parker 1962:1.
5. Blackman 1925:252.
6. Frankfurter 1998:145.
7. Sauneron 2000:97.
8. Bohleke 1997:159.
9. Frankfurter 1998:154.
10. Frankfurter 1998:146.
11. Sauneron 2000:102.
12. Frankfurter 1998a:149.
13. Sauneron 2000:102.
14. Frankfurter 1998:153.
15. Bohleke 1997:165.
16. Ray 1972:251.
17. Bohleke 1997:158.
18. Frankfurter 1998:51.
19. Foertmeyer 1989:22.
20. Frankfurter 1998a:159.
21. Frankfurter 1998a:158.
22. Foertmeyer 1989:13.
23. Foertmeyer 1989:22.
24. Foertmeyer 1989:69.
25. Sauneron 2000:99.
26. Sauneron 2000:99.
27. Frankfurter 1998a:151.
28. Frankfurter 1998a:151.
29. Frankfurter 1998a:150.
30. McDowell 1990:111.
31. David 2001:282.
32. Bohleke 1997:163.
33. Foertmeyer 1989:80.
34. Frankfurter 1998a:182.
35. Beard 1991:53.
36. McDowell 1999:110.
37. Frankfurter 1998:157.

38. Černý 1962:40.
39. Frankfurter 1998a:146.
40. McDowell 1990:133.
41. McDowell 1990:113.
42. David 2001:280.
43. McDowell 1999:107 and McDowell 1990:108.
44. Blackman 1926:182–3.
45. McDowell 1990:119.
46. McDowell 1990:114.
47. McDowell 1990:117.
48. Dorman 1988:22.
49. Herodotus 1972:181–2.
50. Herodotus 1972:191.
51. Klotz 2010:250.
52. Klotz 2010:251.
53. Černý 1962:36.
54. Frankfurter 1998a:164.
55. Frankfurter 1998a:105.
56. Frankfurter 1998a:64–5.
57. Frankfurter 1998a:155.
58. Frankfurter 1998a:162.
59. Frankfurter 1998:165.
60. Sauneron 2000:103.
61. Strabo, Book XVII.
62. David 2001:316.
63. Bohleke 1997:162.
64. Klotz 2010:252.
65. Černý 1962:35.
66. Frankfurter 1998:159.
67. Bohleke 1997:159.
68. Frankfurter 1998a:160.
69. Frankfurter 1998a:108.
70. Frankfurter 1998a:108.
71. Foertmeyer 1989:74.
72. Frankfurter 1998a:166.
73. Frankfurter 1998:35.
74. Frankfurter 1998a:193.
75. Frankfurter 1998a:194.
76. Frankfurter 1998a:194.
77. Frankfurter 1998a:197.
78. Frankfurter 1998a:161.
79. Frankfurter 1998a:152.
80. Frankfurter 1998a:178.
81. Ripat, 2006:310.
82. Ripat, 2006:311.
83. Ripat, 2006:312.
84. Frankfurter 1998a:191.
85. Frankfurter 1998a:157 and P. Lond, III, 854.
86. David 2001:319.
87. Strabo 17.1.43.
88. Foertmeyer 1989:309.
89. Frankfurter 1998:18.
90. Landvatter 2019:154.
91. Landvatter 2019:164.
92. Landvatter 2019:164.

93. Landvatter 2019:164.
94. Foertmeyer 1989:70.
95. Grafton Milne 1916:79.
96. Frankfurter 1998a:171.
97. Frankfurter 1998a:170.
98. Frankfurter 1998a:170.
99. Foertmeyer 1989:75.
100. Frankfurter 1998a:174.
101. Frankfurter 1998a:148.
102. Ryholt 1993:192.
103. McDowell 1990:109.

**Chapter 8**
1. Grafton Milne 1916:76.
2. McDowell 1999:157.
3. Lichtheim 1976:189. 'Sister' was a term of endearment and often applied to a lover rather than a sibling.
4. Lichtheim 1973:170.
5. In a non-monetary society goods were used for purchasing power and wages were paid in set rations of food.
6. Goedicke 1984:18.
7. Goedicke 1984:78.
8. Goedicke 1984:121.
9. Goedicke 1984:43.
10. Lichtheim 1976:191.
11. Grafton Milne 1916:76.
12. Coleman Darnell & Manassa 2007:188 & 201.
13. Coleman Darnell & Manassa 2007:201.
14. Partridge 2002:98.
15. Coleman Darnell & Manassa 2007:204.
16. Lichtheim 1976:226.
17. Habachi 1958:15.
18. Habachi 1958:20.
19. Tyldesley 2005:22.
20. Cooney 2007:166.
21. Hawass 2006:158.
22. Cooney 2007:166.
23. Frankfurter 1998a:218.
24. Rutherford 2006:135.
25. Rutherford 2006:138.
26. Breasted 1906.
27. Rutherford 2003:173.
28. Landvatter 2019:156.
29. Landvatter 2019:156.
30. Rutherford 2003:174.
31. Müller 2019:228.
32. Adams 2019:58.
33. Lichtheim 1988:101.
34. Eleventh dynasty stela of Nakhty in Snape 2019:259.
35. Adams 2019:60.
36. Müller 2019:235–7.
37. Kelly Simpson 1974:12.
38. Stela of Intef-Iker Year 33 of Senwosret I (1971–1926 BCE) in Snape 2019:259.
39. Kelly Simpson 1974:6.
40. O'Connor 1985:170.

41. Snape 2019:256.
42. Lichtheim 1988:103.
43. Lichtheim 1988:122–3.
44. Adams 2019:66.
45. Rutherford 2003:173.
46. Müller 2019:239.
47. Herodotus 1972:152–3.
48. Rutherford 2006:143.
49. Rutherford 2006:136.
50. Marković 2017:146.
51. Pinch 1993:251–3.
52. Herodotus 1972:65–7.
53. Rutherford 2006:146.
54. Rutherford 2006:144.

**Epilogue**
1. Tomazos 2017.
2. Tomazos 2017.
3. Tomazos 2017.
4. Tomazos 2017.
5. El-Sayad El Naggar 2013:2.
6. Cook 2014.
7. Tomazos 2017.
8. Hanna 2013:374.
9. Hanna 2013:371.
10. Hanna 2013:372.
11. Parcak 2015:196.
12. Parcak 2015:197.
13. Parcak 2015:199.
14. Hanna 2013:374.
15. Parcak 2015:202.
16. Hanna 2013:375.
17. Hanna 2013:374.
18. Tomazos 2017.
19. Butler & Suntikul 2013:3.
20. Tomazos 2017.
21. Gunaratna 2015:106.
22. Gunaratna 2015:108.
23. https://www.reuters.com/article/us-egypt-regeni-exclusive-idUSKCN0XI1YU (accessed 27 October 2022).
24. https://www.bbc.co.uk/news/world-middle-east-34241680 (accessed 27 October 2022).
25. Tomazos 2017.
26. Tomazos 2017.
27. El-Sayad El Naggar 2013:1.
28. Nemoy 1939:27.

# References

Abu-Zekry, T. and A. El-Kholei. 1998. *Tourism and Tourists in the Built Environment of Egypt in the Age of Globalization Conference Paper: The International Association for the Study of Traditional Environments (IASTE)*.

Abdulrahman, Z. 2021 *Egyptian Bellydancer, A Disappearing Act*. https://www.zaraszouk.com/product-page/lecture-egyptian-dancers-a-disappearing-act (Accessed 23 March 2023).

Adams, M. 2019. 'The Origins of Sacredness at Abydos'. In *Abydos: The Sacred Land at the Western Horizon*. Edited by I. Regulski, 25–70. London: Peeters.

Anderson, R. and I. Fawsy (eds). 1987. *Egypt in 1800*. London: IMPADS Associates.

Bagnall, R.S. and R. Cribiore. 2006. *Women's Letters from Ancient Egypt, 300 BC–AD 800*. Ann Arbor: University of Michigan Press.

Baines, J. and J. Malek. 1984. *Atlas of Ancient Egypt*. Oxford: Facts on File.

Bagnall, R.S. and D.W. Rathbone. 2004. *Egypt from Alexander to the Copts*. London: British Museum Press.

Barnes, R. 2004. 'Cloistered Bookworms in the Chicken Coop of the Muses.' In *The Library of Alexandria: Centre of Learning in the Ancient World*. Edited by R. Macleod, 61–78. London: I.B.Tauris.

Beard, M. 1991. *Literacy in the Roman World*. Ann Arbour: University of Michigan Press.

Bednarski, A. 2005. *Holding Egypt: Tracing the Reception of the 'Description de L'Egypte' in Nineteenth-century Great Britain*. London: Golden House.

Behbehanian, L. 2000. 'Policing the Illicit Peripheries of Egypt's Tourism Industry'. *Middle East Report*, no. 216, 32–34.

Behlmer, H. 1998. 'Visitors to Shenoute's Monastery'. In *Pilgrimage and Holy Space in Late Antique Egypt*. Edited by D. Frankfurter, 341–71. London: Brill.

Beness, J.L. & T. Hillard. 2003. 'The First Roman at Philae ('CIL' 1.2².2937a)'. *Zeitschrift für Papyrologie und Epigraphik* 144, 203–07.

Blackman, A.M. 1925. 'Oracles in Ancient Egypt'. *The Journal of Egyptian Archaeology* 11, no. 3/4, 249–55.

Blackman, A.M. 1926. 'Oracles in Ancient Egypt II'. *The Journal of Egyptian Archaeology* 12, no. 3/4, 176–85.

Bohleke, B. 1997. 'An Oracular Amuletic Decree of Khonsu in the Cleveland Museum of Art'. *The Journal of Egyptian Archaeology* 83, 155–67.

Boorstin, D. 1961: *The Image: A Guide to Pseudo-Events in America*. New York: Vintage Books.

Booth, C. 2018. *Excavating Paper Squeezes: Identifying the Value of Nineteenth and Early Twentieth Century Squeezes of Ancient Egyptian Monuments, Through the Collections of Seven UK Archives*. Unpublished thesis held at University of Birmingham. https://etheses.bham.ac.uk/id/eprint/8715/1/Booth18PhD.pdf.

Brazil, W. 2000. 'Alexandria: The Umbilicus of the Ancient World'. In *The Library of Alexandria: Centre of Learning in the Ancient World*. Edited by R. MacLeod, 35–59. London: I.B. Tauris.

Brendon, P. 1991. *Thomas Cook: 150 Years of Popular Tourism*. London: Secker & Warburg.

Breasted, H.J. 1906. *Ancient Records of Egypt*. Volume I. Chicago: University of Chicago Press.

Broadhurst, J.C. (trans). 1952. *The Travels of Ibn Jubayr*. London: Goodword Books.

Bruce, J. 1790. *Travels to Discover the Source of the Nile*. Edinburgh.

Budge, E.A.W. 1906. *Cook's Handbook for Egypt and the Egyptian Sudan*. London: Thomas Cook.

Burnett, C. 2003. 'Images of Ancient Egypt in the Latin Middle Ages'. In *The Wisdom of Egypt*. Edited by P. Ucko and T. Champion, 65–99. London: University College London Press.

Butler R. and W. Suntikul (eds). 2013. *Tourism and Political Change*. Oxford: Goodfellow.

Buzard, J. 1993. *The Beaten Path: European Tourism, Literature, and the Ways to Culture, 1800–1918*. Oxford: Oxford University Press.
Carnarvon, F. 2009. *Egypt at Highclere: The Discovery of Tutankhamun*. Newbury: Highclere Enterprises.
Černý, J. 1962. 'Egyptian Oracles'. In *A Saite Oracle Papyrus From Thebes in the Brooklyn Museum, Rhode Island*. Edited by R.A. Parker, 35–48. Providence: Brown University Press.
Chauveau, M. 2002: *Cleopatra: Beyond the Myth*. Ithaca: Cornell University Press.
Coleman Darnell, J. and C. Manassa. 2007. *Tutankhamun's Armies: Battle and Conquest During Ancient Egypt's Late Eighteenth Dynasty*. London: John Wiley & Sons.
Cook, S. 2014. 'Egypt's Solvency Crisis in Council on Foreign Relations: Contingency Planning Memorandum No 20'. https://www.cfr.org/report/egypts-solvency-crisis (Accessed 14 February 2023).
Cooney, K. 2007. 'Labour'. In *The Egyptian World*. Edited by T. Wilkinson, 160–74. London: Routledge.
Curran, B.A. 2003. 'The Renaissance Afterlife of Ancient Egypt (1400–1650)'. In *The Wisdom of Egypt*. Edited by P. Ucko and T. Champion, 101–31. London: University College London Press.
Dannenfeldt, K.H. 1959. 'Egypt and Egyptian Antiquities in the Renaissance'. *Studies in the Renaissance* 6: 7–27.
David, R. 2002. *Religion and Magic in Ancient Egypt*. London: Penguin.
Davis, S.J. 1998. 'Pilgrimage and the Cult of Saint Theclas in Late Antique Egypt in Frankfurter'. In *Pilgrimage and Holy Space in Late Antique Egypt*. Edited by D. Frankfurter, 303–39. London: Brill.
Dorman, P.F. 1988. *The Monuments of Senenmut: Problems in Historical Methodology*. New York: Kegan Paul.
Edwards, A. 1878. *A Thousand Miles Up the Nile*. Leipzig: Bernhard Tauchnitz.
El-Kholei A. and T. Abu Zekry. 1998. 'Tourists in the Built Environment of Egypt in the Age of Globalization'. https://www.researchgate.net/publication/299584964_Tourism_and_Tourists_in_the_Built_Environment_of_Egypt_in_the_Age_of_Globalization (Accessed 26 August 2022).
El Aref, N. 2006. 'Under the Waves'. *Al Ahram Weekly*, republished https://english.ahram.org.eg/NewsContent/9/40/354/Heritage/Ancient-Egypt/Under-the-waves.aspx (Accessed 18 August 2022).
El-Bastawissi, I. 2000. 'Need for a Coastal Management Plan: The Alexandria Case'. In *Underwater Archaeology and Coastal Management: Focus Alexandria*. Edited by M.H. Mostafa, N. Grimal, and D. Nakashima, pp.164–178 Paris: UNESCO Publishing.
El Daly, O. 2003. 'Ancient Egypt in Medieval Arabic Writings'. In *The Wisdom of Egypt*. Edited by P. Ucko and T. Champion, 39–63. London: University College London Press.
El Daly, O. 2003a. *Ancient Egypt in Medieval Moslem/Arabic Writings*. London: Unpublished PhD Thesis. University College London.
El Daly, O. 2005. *Egyptology: The Missing Millennium*. Ancient Egypt in Medieval Arabic Writings. London: University College London Press.
El-Gawhary, K. 1995. 'Sex Tourism in Cairo'. *Middle East Report: Sept-Oct No. 196: Tourism and the Business of Pleasure*, 26–27.
El Sayed El-Naggar, A. 2013. 'The Post-Arab Spring Economic Outlook in Egypt'. *Asian Institute for Policy Studies*, no. 43, 1–7.
Flaubert, G. 1979. *Flaubert in Egypt*. Translated and edited by Francis Steegmuller. London: Penguin.
Foertmeyer, V.A. 1989. *Tourism in Graeco-Roman Egypt*. Unpublished PhD for Princeton University Department of Classics. https://catalog.princeton.edu/catalog/995111743506421 (Accessed 23 March 2023).
Francaviglia, R.V. 2011. *Go East, Young Man: Imagining the American West as the Orient*. Denver: University Press of Colorado.
Frankfurter, D. 1998. 'Introduction: Approaches to Coptic Pilgrimage'. In *Pilgrimage and Holy Space in Late Antique Egypt*. Edited by D. Frankfurter, 3–48. London: Brill.
Frankfurter, D. 1998a. *Religion in Roman Egypt Assimilation and Resistance*. Princeton: Princeton University Press.

Frankfurter, D. 2006. 'Urban Shrine and Rural Saint in Fifth Century Alexandria'. In *Pilgrimage in Graeco-Roman and Early Christian Antiquity: Seeing the Gods*. Edited by J. Elsner et al, 435–99. Oxford: Oxford University Press.

Gabra, G. 1993. *Cairo: The Coptic Museum and Old Churches*. Cairo: Egyptian International Publishing.

Gange, D. 2015. 'The Ruins of Preservation: Conserving Ancient Egypt 1880–1914'. *Past and Present* 226 (suppl 10), 78–99.

Goedicke, H. 1984. *Studies in the Hekanakhte Papers*. Baltimore: Halgo.

Grafton Milne, J. 1916. 'Greek and Roman Tourists in Egypt'. *The Journal of Egyptian Archaeology* 3, 76–80.

Gray, M. 1998. 'Economic Reform, Privatization and Tourism in Egypt'. *Middle Eastern Studies* 34, no 2, 91–112.

Grossman, P. 1998. 'The Pilgrimage Center of Anu Mina'. In *Pilgrimage and Holy Space in Late Antique Egypt*. Edited by D. Frankfurter, 282–302. London: Brill.

Guglielmi, W. 2001. 'Agatha Christie and her Use of Ancient Egyptian Sources'. In *Agatha Christie and Archaeology*. Edited by C. Trümpler, pp.351–90. London: British Museum Press.

Gunaratna, R. 2015. *Egypt in Counter Terrorist Trends and Analyses* 8, no 1, 106–11.

Habachi, L. 1958. 'The Graffiti and Work of the Viceroys of Kush in the Region of Aswan'. *Kush* 6, 13–36.

Hamed, A. 2013. 'Re-Excavation of Seti First Tomb, KV17, Luxor, Egypt'. *International Journal of Conservation Science* 4, no. 4, 433–46.

Hamilton, J. 2005. *Thomas Cook: The Holiday Maker*. Stroud: Sutton Publishing.

Hankey, J. 2007. *Passion for Egypt: Arthur Weigall, Tutankhamun and the 'Curse of the Pharaohs'*. Barbara Ward and Associates. London.

Hanna, M. 2013. 'What Has Happened to Egyptian Heritage after the 2011 Unfinished Revolution?' *Journal of Eastern Mediterranean Archaeology and Heritage Studies* 1, no. 4, 371–75.

Hawass, Z. 2006. *Mountains of the Pharaohs*. Cairo: American University in Cairo Press.

Haycock, D.B. 2003. 'Ancient Egypt in 17th and 18th-century England'. In *The Wisdom of Egypt*. Edited by P. Ucko and T. Champion, 134–60. London: University College London Press.

Herodotus. *The Histories*. Translated by A. de Sélincourt. London: Penguin (1972).

Hopwood, D. 1987. *Egypt: Politics and Society, 1945–1990*. London: Routledge.

Humphreys, A. 2011. *Grand Hotels of Egypt in the Golden Age of Travel*. Cairo: American University in Cairo Press.

Humphreys, A. 2021. *On The Nile: In the Golden Age of Travel*. Cairo. American University in Cairo Press.

Hunter, F.R. 2004. 'Tourism and Empire: The Thomas Cook and Son Enterprise on the Nile 1868–1914'. *Middle Eastern Studies* 40, 28–54.

Ibrahim, M.A. and M.A. Ibrahim. 2011. 'The Determinants of International Tourist Demand for Egypt: Panel Data Evidence'. In *European Journal of Economics, Finance and Administrative Sciences* 30, 50–58.

Jackson, K. and J. Stamp. 2002. *Pyramid: Beyond Imagination*. London. BBC Worldwide.

Jensen, M.N. 1990. 'Homeric Scholarship'. In *Alexandria; a Cultural and Religious Melting Pot*. Edited by G. Hinge G. and J. Krasilnikoff. Aarhus: Aarhus University Press.

Jones, M. 2003. 'The Work of the American Research Centre in Egypt in the Tomb of Sety I in the Valley of the Kings, 1998–1999'. In *Egyptology at the Dawn of the Twenty-first Century, Volume 1*. Edited by Z. Hawass, 252–61. Cairo: The American University in Cairo Press.

Keck, S. 2010. '"Going Out and Doing Something": Victorian Tourists in Egypt and the "Tourist Ethic"'. *Journal of Tourism and Cultural Change* 8, no. 4, 293–304.

Kelly Simpson, W. 1974. *The Terrace of the Great God at Abydos: The Offering Chapels of Dynasties 12 and 13*. New Haven: Peabody Museum of Natural History of Yale University.

Klotz, D. 2010. 'Two Overlooked Oracles'. *The Journal of Egyptian Archaeology* 96, 247–54.

Kuppinger, P. 2005. 'Globalization and Exterritoriality in Metropolitan Cairo'. *Geographical Review* 95, no. 3, *New Geographies of the Middle East*, 348–72.

Landvatter, T. 2019. 'Fluctuating Landscapes: Cross-cultural Interaction, Mortuary Practice and Ritual Art at Ptolemaic-Roman Abydos'. In *Abydos: The Sacred Land at The Western Horizon*. Edited by I. Regulski, 153–69. London: Peeters.

Lichtheim, M. 1973. *Ancient Egyptian Literature: Volume I the Old and Middle Kingdoms.* Berkeley: University of California Press.
Lichtheim, M. 1976. *Ancient Egyptian Literature: Volume II the New Kingdom.* Berkeley: University of California Press.
Lichtheim, M. 1988. *Ancient Egyptian Autobiographies.* Fribourg: Biblical Institute of the University of Fribourg.
Lloyd, A. 2000. 'The Ptolemaic Period (332–30 BC)'. In *The Oxford History of Ancient Egypt.* Edited by I. Shaw, 388–413. Oxford: Oxford University Press.
Macquitty, W. 1965. *Abu Simbel.* London: Macdonald.
Maehler, H. 2004. 'Alexandria: The Mouseion, and Cultural Identity'. In *Alexandria, Real and Imagined.* Edited by A. Hirst and M. Silk, 1–14. London: Routledge.
Mairs, R. and M. Muratov. 2015. *Archaeologists, Tourists, Interpreters: Exploring Egypt and the Near East in the Late 19$^{th}$–Early 20$^{th}$ Centuries.* London: Bloomsbury Academic.
Mandler, P. 1999. '"The Wand of Fancy": The Historical Imagination of the Victorian Tourist'. In *Material Memories.* Edited by M. Kwint, et al., 125–41. Oxford: Berg Publishers.
Manley, D. 1991. *The Nile: A Traveller's Anthology.* London: Cassell.
Marković, N. 2017. 'The Majesty of Apis has Gone to Heaven: Burial of the Sacred Apis Bull within the Landscape of Memphis During the Late Period (664–332 BCE)'. In *Burial and Mortuary Practices in Late Period and Graeco-Roman Egypt.* Edited by K. Kóthay, 145–54. Budapest: Museum of Fine Arts.
McDowell, A.G. 1990. *Jurisdiction in the Workmen's Community of Deir el Medina.* Leiden: Nederlands instituut Voor Het Nabije Oosten.
McDowell, A.G. 1999. *Village Life in Ancient Egypt.* Oxford: Oxford University Press.
McKenzie, J. 2003. 'Glimpsing Alexandria from Archaeological Evidence'. *Journal of Roman Archaeology* 16, 35–63.
Meinardus, O. 2002. *Coptic Saints and Pilgrimages.* Cairo: The American University Press in Cairo.
Mikhail, M.S. and T. Vivian. 1997. 'Life of Saint John the Little: An Encomium by Zacharias of Sakha'. *Coptic Church Review* 18, nos. 1–2, 17–64.
Mitchell, T. 1995. 'Worlds Apart: An Egyptian Village and the International Tourist Industry'. *Middle East Report Sep-Oct.*, 8–11 and 23.
Momigliano, A. 1966. *Studies in Historiography.* London: Weidenfeld Goldbacks.
Montserrat, D. 1998. 'Pilgrimage to the Shrine of SS Cyrus and John at Menouthis in Late Antiquity'. In *Pilgrimage and Holy Space in Late Antique Egypt.* Edited by D. Frankfurter, 257–79. London: Brill.
Morcos, S., N. Tongring, Y. Halim, M. El-Abbaddi and H. Awad. 2003. *Towards Integrated Management of Alexandria's Coastal Heritage.* Alexandria: Coastal Region and Small Island, Paper 14.
Morcos, S.A. 2000. 'Early Discoveries of Submarine Archaeological Sites in Alexandria'. In *Underwater Archaeology and Coastal Management: Focus on Alexandria.* Edited by M.H. Mostafa et al., 33–45. Paris: UNESCO Publishing.
Morecroft, A.S. 2018. *The Enlightenment Rediscovery of Egyptology: Vitaliano Donati's Egyptian Expedition, 1759–62.* London: Routledge.
Müller, V. 2019. 'A History of Millennia: The Configerate of a Cultic Landscape Around the Tomb of King Den at Umm el-Qa'ab, Abydos'. In *Abydos: The Sacred Land at The Western Horizon.* Edited by I. Regulski, 215–34. London: Peeters.
Nemoy, L. 1939. 'The Treatise on the Egyptian Pyramids (Tuhfut al-kiram fi khabar al-ahram)'. *Isis* 30, no. 1, 17–37.
Neret, G. 2002. *Description de L'Egypte.* London: Taschen.
O'Connor, D. 1985. 'The Cenotaphs of the Middle Kingdom at Abydos'. In *Mélanges Gamal Eddin Mokhtar Vol II.* Edited by P. Posener-Kriéger, 161–77. Cairo: Institut Français d'archéologie orientale du Caire.
Parcak, S. 2015. 'Archaeological Looting in Egypt: A Geospatial View (Case Studies From Saqqara, Lisht, and el Hibeh)'. *Near Eastern Archaeology* 78, no. 3, 196–203.
Parker, R.A. 1962. *A Saite Oracle Papyrus From Thebes in the Brooklyn Museum.* Providence: Brown University Press.

Partridge, R. 1996. *Transport in Ancient Egypt*. London: Rubicon.
Partridge, R. 2002. *Fighting Pharaohs: Weapons and Warfare in Ancient Egypt*. Manchester: Peartree Publishing.
Pinch, G. 1993. *Votive Offerings to Hathor*. Oxford: Griffith Institute.
Plutarch. 1973. *The Age of Alexander: Nine Greek Lives by Plutarch*. Translated by I. Scott-Kilvert. London: Penguin.
Ray, J. 1972. 'Two Inscribed Objects in the Fitzwilliam Museum, Cambridge'. *The Journal of Egyptian Archaeology* 58, 247–53.
Reeves, N. and R. Wilkinson. 1996. *The Complete Valley of the Kings*. London: Thames and Hudson.
Richter, T. and C. Steiner. 2008. 'Politics, Economics and Tourism Development in Egypt: Insights into the Sectoral Transformations of a Neo-Patrimonial Rentier State'. *Third World Quarterly* 29 no. 5, 939–59.
Ripat, P. 2006. 'The Language of Oracular Inquiry in Roman Egypt'. *Phoenix* 60, no. 3/4, 304–28.
Rodenbeck, M, 1988. *Cairo: The City Victorious*. London: Picador.
Romer, J. 1981. *Valley of the Kings*. London: Michael O'Mara Books.
Russell, T. 2001. *The Napoleonic Survey of Egypt, Description de l'Égypt: The Monuments and Customs of Egypt; Selected Engravings and Texts*. Aldershot: Ashgate Publishing.
Rutherford, I. 1998. 'Island of the Extremity: Space, Language and Power in the Pilgrimage Traditions of Philae'. In *Pilgrimage and Holy Space in Late Antique Egypt*. Edited by D. Frankfurter, 230–56. London. Brill.
Rutherford, I. 2003. 'Pilgrimage in Graeco-Roman Egypt'. In *Ancient Perspectives on Egypt*. Edited by R. Matthews and C. Roemer, 171–89. New York: Routledge.
Rutherford, I. 2006. 'Down-Stream to the Cat-Goddess: Herodotus on Egyptian Pilgrimage'. In *Pilgrimage in Graeco-Roman and Early Christian Antiquity: Seeing the Gods*. Edited by J. Elsner et al, 131–48. Oxford: Oxford University Press.
Ryholt, K. 1993. 'A Pair of Oracle Petitions Addressed to Horus-of-the-Camp'. *The Journal of Egyptian Archaeology* 79, 189–98.
Sauneron, S. 2000. *The Priests of Ancient Egypt*. London: Cornell University Press.
Slyomovics, S. 1989. 'Cross-Cultural Dress and Tourist Performances in Egypt'. *Performing Arts Journal* 11, no. 3–12, no 1. The Intercultural Issue, 139–48.
Snape, S. 2019. 'Memorial Monument at Abydos and "The Terrace of the Great God"'. In *Abydos: The Sacred Land at the Western Horizon*. Edited by I. Regulski I, 255–72. London: Peeters.
Thompson, J. 2015. *Wonderful Things: A History of Egyptology, 1: From Antiquity to 1881*. Cairo: The American University in Cairo Press.
Timbie, J. 1998. 'A Liturgical Procession in the Desert of Apa Shenoute'. In *Pilgrimage and Holy Space in Late Antique Egypt*. Edited by D. Frankfurter, 415–41. London: Brill.
Tomazos, K. 2017. 'Egypt's Tourism Industry and the Arab Spring'. In *Tourism and Political Change*. Edited by R. Butler and W. Suntikul (2nd ed.). https://pdfs.semanticscholar.org/96f1/7a6df398029386fac23c2044b3f0e474a45f.pdf (Accessed 27 October 2022).
Tyldesley, J. 2005. *Egypt: How a Lost Civilisation was Rediscovered*. London: BBC Books.
Tyldesley, J. 2009. *Cleopatra: Last Queen of Egypt*. London: Profile Books.
Usick, P. 2002. 'Berths Under the Highest Stars: Henry William Beechey in Egypt 1816–1819'. In *Egypt through the Eyes of Travellers*. Edited by P. Starkey and N. El Kholy, 13–24. London: Astene.
Vallance, J. 2000. 'Doctors in the Library: The Strange Tale of Apollonius the Bookworm and Other Stories'. In *The Library of Alexandria: Centre of Learning in the Ancient World*. Edited by R. MacLeod, 95–113. London: I.B. Tauris Publishers.
Wynne-Hughes, E. 2012. '"Who Would Go to Egypt?": How Tourism Accounts for "Terrorism"'. *Review of International Studies* 38, no 3, 615–40.
Zahran, M. 2000. 'Urban Design and Eco-Tourism: The Alexandria Comprehensive Master Plan'. In *Underwater Archaeology and Coastal Management: Focus Alexandria*. Edited by M.H. Mostafa M. H., N. Grimal and D. Nakashima, 185–89. Paris: UNESCO Publishing.
Zallio, F. 2010. *Egypt after the Crisis: Resilience and New Challenges*. German Marshall Fund of the United States, May 2010.

# Index

Abu Mina 72, 84, 85, 86
Abu Simbel ix, 5, 8, 27, 61, 62, 63, 133. 164, 184
Abu-al Haggag Mosque 104
Abydos 30, 33, 74, 109, 117, 124, 15, 153, 15, 169, 170, 171, 172, 173
Akhmim 68, 78, 97, 98, 99, 100, 104
Alexander the Great xviii, 89, 107, 110, 111, 119, 120, 124, 133, 154
Alexandria xviii, 3-7, 33, 38, 41, 43, 51, 55-57, 66, 68, 70, 72-74, 77, 79, 80, 81, 83, 84, 88, 89, 95-97, 105, 107, 109, 110- 122, 132, 140, 142, 148-150, 152, 155, 172
Al-Sisi, Abdul Fattah 182, 183
al-Suyūṭī, Jalāl al-Dīn 90, 91, 100, 102, 103, 184
Amarna 124, 165, 181
Amenhotep I 134, 140, 144, 150
Amenhotep II 133, 166
Amenhotep III xii, 105, 125, 128, 166, 177, 179
Amun 111, 115, 123, 135, 136, 145, 147, 150, 154, 158, 166, 174, 176, 195 n4
Ancestor bust 174
Antinoopolis xvi, 60, 113, 115, 124, 173
Apis bull 110, 111, 115, 116, 119, 125, 149, 150, 175, 176
Arsinoë 60, 112, 116

Babylon 68, 74, 76, 80, 82, 83, 119
Belly dancing 20-21, 24
Belzoni, Giovanni x, 18, 28, 32, 60-63
Bes 125, 153, 155-156, 173
Buchis bull 118, 150

Caesar, Augustus 111
Caesar, Julius 105, 111, 117, 118, 119, 122, 123
Caesar, Octavian 118
Cairo 15, 16, 19-24, 30, 32, 33, 36, 38, 39, 40, 41, 42, 43, 45, 46, 47, 49, 51, 56, 64, 66, 68, 69, 73, 74, 76, 82, 833, 84, 85, 88, 89, 90, 92, 93, 97, 98, 102, 103, 106, 141, 173, 180, 181, 183, 184
Canopus 82, 115, 116, 149, 152
Cleopatra xviii, 39, 88, 117, 118, 109, 122, 123

Colossi of Memnon 33, 63, 68, 105, 114, 115, 116, 117, 118, 125, 126, 128, 129, 133
Cruise xvi, 1, 2, 3, 10, 11, 19, 26, 30, 33, 38, 39, 42, 47, 52, 112, 117, 118

Dahab 19, 23, 24, 36
Dahebeya xvi, 19, 28, 30, 32, 33, 36, 37, 40, 42
Dashur 181
Deir el Bahri xvi, 3, 9, 32, 117, 139, 140, 174, 176, 177, 179
Deir el Medina 9, 12, 134, 136, 137, 140, 143, 144, 145, 158
Dendera 30, 33, 57, 68, 95, 97, 104, 142, 174, 177
Description de l'Égypte 55, 58
Donkey xii, xiii, 33, 35, 38, 48, 50, 53, 69, 81, 114, 115, 140, 159, 160
Dr Raghab's Pharaonic Village 1, 2

Edfu 8, 32, 38, 62, 174, 175
Edwards, Amelia xi, 26, 27, 36, 37, 40, 41, 44
El-Gourna 14

Festival, Beautiful Festival of the Valley 174
Festival, Khoiak 132, 175
Festival, Opet Festival 165, 174
Festival, yq Festival 131

Garbage City 21, 22
Giza ix, xiv, xv, 1, 2, 8, 9, 12, 15, 16, 18, 20, 33, 34, 35, 46, 47, 51, 59, 61, 64, 68, 69, 87, 88, 90, 91, 98, 100, 102, 103, 106, 112, 115, 123, 145, 167, 180, 186
Graffiti viii, ix, xvii, xviii, xix, 5, 27, 28, 72, 81, 108, 116, 117, 122, 124, 125, 126, 127, 128, 129, 130, 131, 132, 133, 140, 152, 155, 156, 158, 164, 166, 167, 168
Granaries of Joseph 87, 88

Hanging Church 82
Herodotus xiv, xviii, xiv, 101, 107, 108, 109, 121, 123, 124, 146, 173, 175, 177, 179
Holy Family 74, 76, 82, 87
Homer 56, 110, 112, 115, 125, 126, 132
Hurghada 2, 19, 20, 183

Ibn Tulun 91, 95
Isis 59, 69, 82, 88, 113, 118, 124, 125, 130, 131, 132, 140, 144, 147, 148, 149, 152, 156, 172, 178
Imhotep 103, 139, 179

Jesus 71, 74, 75, 77, 88, 102, 129
Joseph 73, 74, 75, 87, 88, 103
Jubayr, Ibn 88, 95, 96, 97, 98, 99, 103, 104

Kalabsha 76, 114, 132, 133, 151
Karnak xii, 1, 5, 8, 17, 32, 64, 67, 69, 73, 150, 157, 165, 174, 175, 176, 183
Knbt 144-145
Kom el Dikka 6, 7

Library of Alexandria 105, 121, 122
Lieder, Alice 30, 32
Lighthouse of Alexandria 95, 96, 97
Luxor xii, xiv, 3, 4, 8, 9, 10, 11, 13, 14, 15, 19, 20, 22, 23, 30, 32, 33, 38, 40, 42, 47, 48, 49, 51, 53, 54, 56, 57, 59, 62, 65, 67, 68, 104, 134, 139, 145, 150, 163, 165, 174, 182, 184

Matariya 74, 75, 88, 102
Medinet Habu xiii, 56, 59, 68, 137
Memnon 62, 63, 116, 117, 118, 125, 126, 128
Memnonium 117, 125, 127, 155, 156
Mena House 43, 46, 47, 49, 51
Menas 84, 85, 86
Morsi, Mohammed 181, 182, 183
Moses 72, 73, 88, 91, 96, 170
Mouseion 120, 121, 122
Mubarak, Hosni 4, 17, 180
Mummies 26, 27, 35, 53, 59, 60, 63, 69, 94, 162, 177, 178
Mut 174, 176

Napoleon 6, 26, 34, 43, 55, 56, 106
Nubia 9, 32, 53, 66, 68, 114, 130, 132, 136, 148, 160

Oracle xix, 72, 77, 82, 84, 104, 111, 115, 125, 132, 133, 134-157, 176
Osiris 69, 110, 125, 132, 148, 155, 156, 169, 170, 171, 172, 173, 175
Oxyrhynchus xvi, 72, 77, 114, 116,124, 173

Papyrus skiff xiv-xv
Pharos Lighthouse 6, 74

Philae xviii, 8, 56, 62, 74, 108, 117, 118, 124, 130, 131, 132, 142, 148, 177
Pompey's Pillar 7, 55, 68, 70, 105, 115
Pyramids xvi, xv, 1, 2, 5, 8, 9, 12, 15, 16, 18, 20, 24, 28, 33, 34, 35, 46, 47, 55, 58, 59, 60, 31, 64, 65, 68, 69, 70, 73, 87, 88, 90, 91, 92, 93, 94, 95, 98, 100, 101, 102, 103, 106, 108, 109, 112, 115, 116, 117, 123-124, 167-168, 171, 179, 181, 184

Relics 72, 74, 78, 79-82, 83, 84, 85, 86, 87, 97, 98, 152

Saint Barbara 82
Saint Cyrus 82, 83, 84, 149, 152
Saint John 82, 83, 84, 149, 152
Saint John the Baptist 79, 80
Saladin 105, 106,
Serapeum 7, 80, 82, 110, 118, 121, 149, 150, 175, 177
Sex tourism 22, 24
Shenoute, Apa 76-79, 81, 84
Shepheard's Hotel 42, 43-45, 53
Souvenirs 28, 37, 52-53, 59, 86, 139, 177
Sphinx 2, 18, 44, 57, 73, 88, 103, 106, 123, 124, 145
Sphinx Avenue xii, 174
Squeezes 30-33, 54, 63
Steamer xvi, 18, 19, 30, 33, 36-39, 41, 43
Strabo 64, 107, 108, 109, 111, 114, 116, 118, 120, 127, 130, 149, 154

Terrorism 3, 179
Thebes 56, 64, 65, 108, 109, 113, 115, 117, 118, 130, 138, 139, 150, 158, 159, 160, 161, 166, 170, 172, 174
Thomas Cook 2, 11, 19, 26, 37-42, 44, 46, 48, 49, 51, 117
Treasure hunting 54, 59, 88, 91-93, 95
Tutankhamun ix, 1, 5, 12, 13, 14, 16, 36, 45, 49, 52, 58, 95

Valley of the Kings 1, 5, 9, 12, 13, 16, 32, 48, 58, 63, 64, 65, 68, 92, 108, 114-117, 126, 127, 128,129, 134, 164

White Monastery 76, 77, 81
Winter Palace 43, 48, 49